applications of geographic information systems

June 2000

This book celebrates the geographic information system (GIS) contributions and accomplishments of ESRI's users during 1999. From Alaska to Australia to Africa, the maps in this book provide a wonderful assortment of cartographic representations. The applications portrayed represent many diverse and rich examples of how GIS is making a difference in the world. Whether it involves looking for trends in breast cancer incidence, devising a new assessment methodology for land parcels, or planning the best route to school, GIS technology is integral to dealing with a range of challenges that the world will face in the 21st century. These maps show how GIS cuts across nearly all disciplines, provides a common language for discussion, and acts as a means to bring people together in the decision making process.

As globalization of businesses and economies brings our world closer, problems increasingly cross international boundaries with widespread implications, and the contributions that GIS users make help facilitate the management of these complex issues.

The future predictions for the application of GIS involve continual growth and development. This will be fueled by a variety of related technologies including wireless access to the Internet, higher data transfer rates, improved remote sensing, and the construction of global spatial databases. With the release of ESRI's recent technology, ArcInfo™ 8 and ArcIMS™ 3, GIS is transitioning to be based on the standard of the information technology community (database management systems, networks, the Web, and open development environments). The long-

term results will be the integration of GIS and location principles in all computers. This will mean increased efficiency and the development of a general-purpose spatial and visual framework for problem solving. The examples shown in this book are proof that this is happening.

Maps have universal appeal, and this year, in order to showcase these beautiful maps to the widest audience, ESRI is publishing portions of the map book on its Web site at www.esri.com/mapmuseum. Many of the contributors are featured in a special biographical section, which includes photos of the map authors.

Warm regards,

Jack Dangermond

contents

Indiana Groundwater Vulnerability Model

Purdue University
West Lafayette, Indiana

*By Mark Ehle, Bernie Engel,
Leighanne Hahn, and Larry Theller*

Contact
Larry Theller
theller@ecn.purdue.edu

Software
ArcView® GIS 3.1 and ArcInfo™ 7.1.2
Hardware
PC and UNIX workstations
Printer
HP DesignJet 2500CP
Data Source(s)
Indiana Geological Survey hydrologic
setting maps, Indiana Department
of Natural Resources, Indiana
Association of Soil and Water
Districts, Indiana Department of
Environmental Management, and
Natural Resources Conservation
Service

Highways
Towns
Pesticide Vulnerability
Low risk
Slight Risk
Medium Risk
High Risk

Pesticide

Several organizations were
involved in the creation of these
maps for the state of Indiana
Ground Water Task Force. The
purpose and scope of the task force are to guide the implementation
of a coordinated, comprehensive, effective groundwater protection and
management program consistent with Indiana's Comprehensive Ground
Water Protection and Management Strategy.

Part of a multifaceted approach to source water protection for Indiana,
this project illustrates a modeled examination of the vulnerability
of generalized aquifer settings to two threat classes. The underlying
hydrologic settings data was created by the Indiana Geological Survey. A
model was constructed of vulnerability of each aquifer's geologic setting
to pesticides or organic chemicals. The vulnerability map was used in
creating management and background sampling areas for the Indiana
State Pesticide Management Plan. A second model was constructed of the
vulnerability to nitrate moving through the root zone to groundwater.

The CD–ROM set of the GIS data is available from the Center for
Advanced Applications in GIS (CAAGIS), Purdue University, ABE 1146,
West Lafayette, Indiana 47907-1146; e-mail caagis@ecn.purdue.edu.

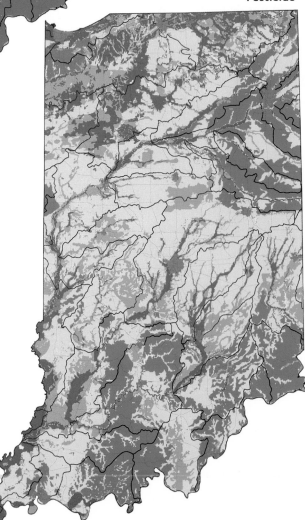

Alternate Harvest Methods

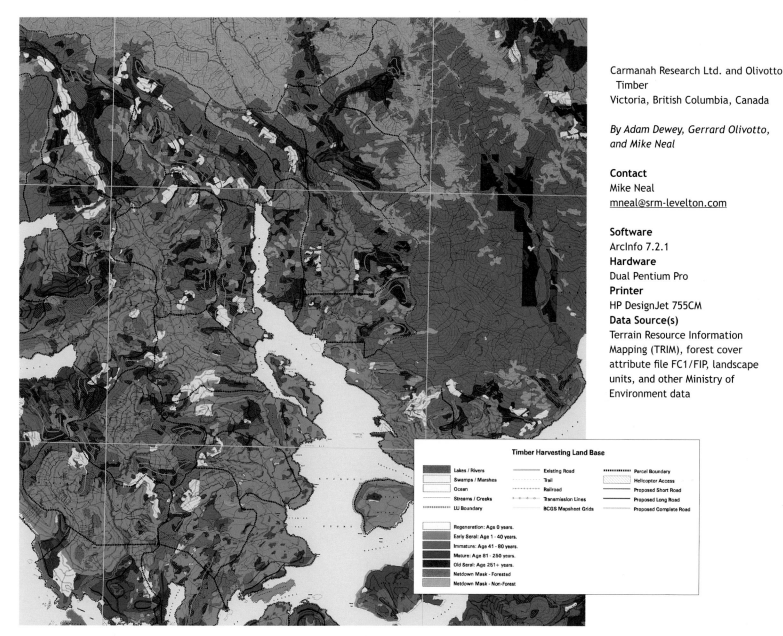

Timber Harvesting Land Base

Lakes / Rivers	Existing Road	Parcel Boundary
Swamps / Marshes	Trail	Helicoptor Access
Ocean	Railroad	Proposed Short Road
Streams / Creeks	Transmission Lines	Proposed Long Road
LU Boundary	BCGS Mapsheet Grids	Proposed Complete Road

Regeneration: Age 0 years.
Early Seral: Age 1 - 40 years.
Immature: Age 41 - 80 years.
Mature: Age 81 - 250 years.
Old Seral: Age 251+ years.
Netdown Mask - Forested
Netdown Mask - Non-Forest

Carmanah Research Ltd. and Olivotto
 Timber
Victoria, British Columbia, Canada

*By Adam Dewey, Gerrard Olivotto,
and Mike Neal*

Contact
Mike Neal
mneal@srm-levelton.com

Software
ArcInfo 7.2.1
Hardware
Dual Pentium Pro
Printer
HP DesignJet 755CM
Data Source(s)
Terrain Resource Information
Mapping (TRIM), forest cover
attribute file FC1/FIP, landscape
units, and other Ministry of
Environment data

This map shows one scenario in a timber harvest cost and availability analysis. A number of scenarios were modeled using an extensive database that includes past harvesting, visually sensitive and recreation areas, forest cover, and terrain. The current (1998) British Columbia Forest Practices code guidelines were used.

The analysis began with the calculation of the timber harvesting land base. Buffers around streams, lakes, and wetlands were designated, and these were merged with areas of steep slope, parks, private lands, and other areas with biological, economic, or social constraints.

Potential road locations and helicopter access areas were digitized, and statistics were produced for timber value, harvest cost, and available volume. Model parameters controlling harvest intensity and methodology were varied, and the model was run to simulate new logging practices.

The results indicated that the concentration of activity in some areas would enable increased protection in other areas, while reducing average harvest costs by approximately 20 percent.

Geographic Distribution of Excess Nitrogen in the Mississippi River Basin

U.S. Department of Agriculture/ARS
National Soil Tilth Lab
Ames, Iowa

By M.R. Burkart and D.E. James

Contact
David James
james@nstl.gov

Software
ArcInfo and ArcView GIS
Hardware
Sun SPARC 20 workstation
Printer
HP DesignJet 755CM
Data Source(s)
1992 Census of Agriculture, State Soil
Geographic (STATSGO) database (U.S.
Department of Agriculture, 1994),
National Atmospheric Deposition
Program/National Trends Network,
and U.S. Geological Survey

Total Sources

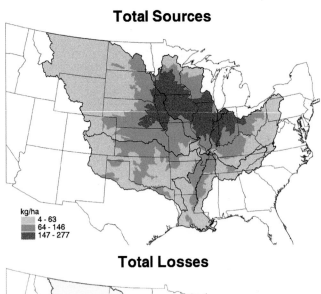

kg/ha
4 - 63
64 - 146
147 - 277

Total Losses

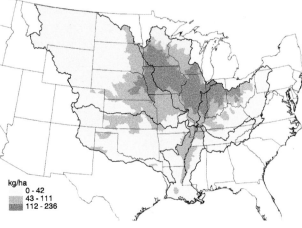

kg/ha
0 - 42
43 - 111
112 - 236

Atmospheric Deposition

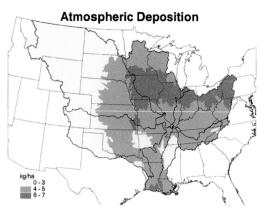

kg/ha
0 - 3
4 - 5
6 - 7

Redeposition of locally derived ammonia

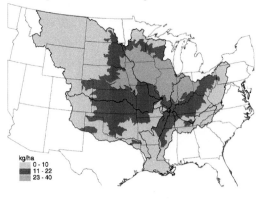

kg/ha
0 - 10
11 - 22
23 - 40

Loss through Crop Senescence

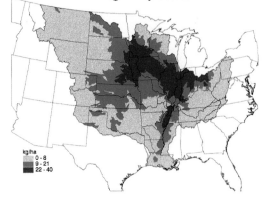

kg/ha
0 - 8
9 - 21
22 - 40

Loss through Crop Harvest

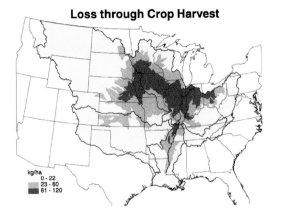

kg/ha
0 - 22
23 - 60
61 - 120

Interest in excess nitrogen has recently been stimulated, in part, by measurements of seasonal hypoxia in the Gulf of Mexico. The cause of the seasonal hypoxia has been indirectly linked to the load of nutrients, particularly nitrogen, delivered to the gulf by rivers, dominantly the Mississippi. Nutrient loads have resulted in nitrogen-limited phytoplankton blooms. Temporal trends of increased biologically bound silica in gulf sediments coincide with increased use of nitrogen fertilizer in the United States. These land use changes and other increases in anthropogenic nitrogen sources contribute to the hypoxia in the gulf.

The purpose of the research, which led to the production of this poster, was to estimate the spatial distribution of several sources and losses of agricultural nitrogen. An important step to understanding the causes and ultimately defining solutions to excess nitrogen is to describe the geographic distribution of nitrogen inputs to the Mississippi River basin. Understanding the geography of nitrogen may contribute to the process of developing the appropriate policy, technology, and education needed to reverse the trend of the increasing load of nitrogen to the basin.

Missouri Digital Soils Survey

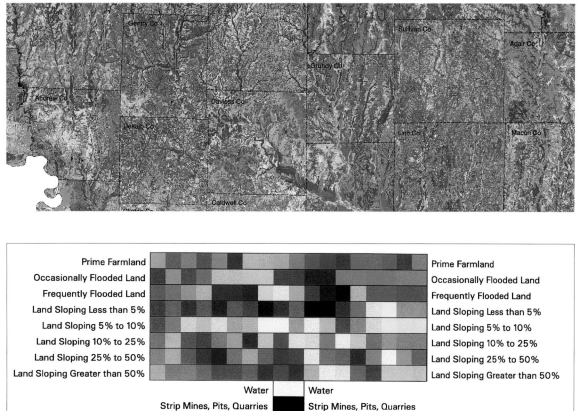

Center for Agricultural, Resource and Environmental Systems (CARES)
University of Missouri-Columbia
Columbia, Missouri

By Scott Burton and Bryan Mayhan

Contact
Scott Burton
burton@cares.missouri.edu

Software
ArcInfo 7.1.2
Hardware
IBM RS6000 workstation
Printer
HP DesignJet 750C+
Data Source(s)
U.S. Department of Agriculture, Natural Resources Conservation Service soil surveys and published and unpublished data

Prime Farmland		Prime Farmland
Occasionally Flooded Land		Occasionally Flooded Land
Frequently Flooded Land		Frequently Flooded Land
Land Sloping Less than 5%		Land Sloping Less than 5%
Land Sloping 5% to 10%		Land Sloping 5% to 10%
Land Sloping 10% to 25%		Land Sloping 10% to 25%
Land Sloping 25% to 50%		Land Sloping 25% to 50%
Land Sloping Greater than 50%		Land Sloping Greater than 50%
Water		Water
Strip Mines, Pits, Quarries		Strip Mines, Pits, Quarries
Flood Control Structures, Dams		Flood Control Structures, Dams
Urban Area		Urban Area

The Missouri Digital Soils Survey map demonstrates the progress of soil survey digitizing in the state of Missouri. Based on the progress of field mapping by soil scientists, it is estimated that digitizing will be completed for the state in the year 2000. The map also graphically represents land use/land cover, topography, and flood hazards within Missouri.

The various soil surveys were scanned, vectorized, and edited using ARC Macro Language (AML™) and ARCEDIT™ software. The soil surveys are assigned lookup table values based on flooding, prime farmland, and slope characteristics for each soil mapping unit. Color classes are assigned for each of the above attributes. Greens indicate prime farmland. Blues show areas that experience frequent flooding, while blue-green sections show areas less often flooded. The yellow-orange-red-violet spectrum follows increasing slopes. Magenta indicates urban areas, while black represents mines and other excavations.

The digital soil layer enables soil scientists, geographers, and other natural resource management personnel to effectively assess geographic patterns. A soil develops under specific climatic regions, vegetative communities, and a geologic parent material. The process is governed by time and influenced by landscape position. GIS technology can be used to identify these patterns. Using this information helps planners become better stewards of the earth.

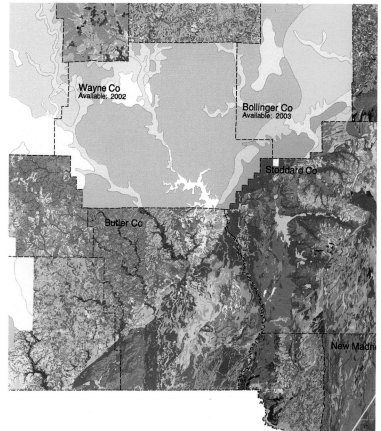

2000 Northwest Region Zone Products

ADVO
Windsor, Connecticut

*By April Cutter, Ron Berthesavage,
and Geographic Data Technology, Inc.*

Contact
GDT Sales
info@gdt1.com

Software
ArcInfo, Maplex, and Adobe
Illustrator
Hardware
HP and Windows NT
Printer
HP DesignJet
Data Source(s)
Dynamap data

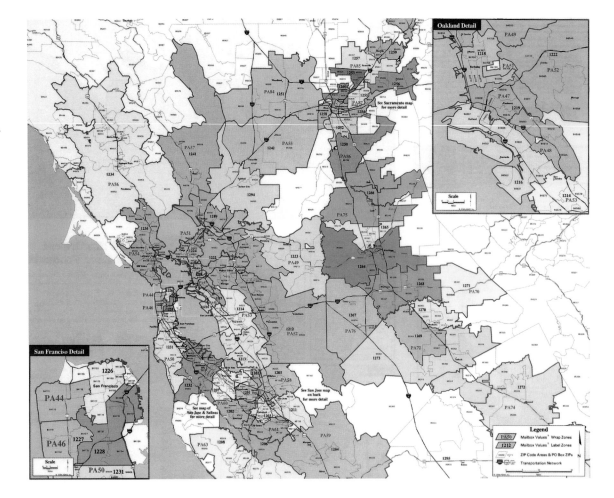

Created by Geographic Data Technology, Inc., for ADVO, a targeted direct mail marketing services company, the Zone Product brochure is a tool that helps describe the ADVO territory to clients. The maps help identify individual neighborhoods and enable clients to clearly see the ADVO coverage area. In this way, clients can have detailed information about where their mail pieces will be delivered. ADVO is the U.S. Postal Service's largest commercial client, distributing 24 billion pieces of mail each year.

Dynamap/Routing

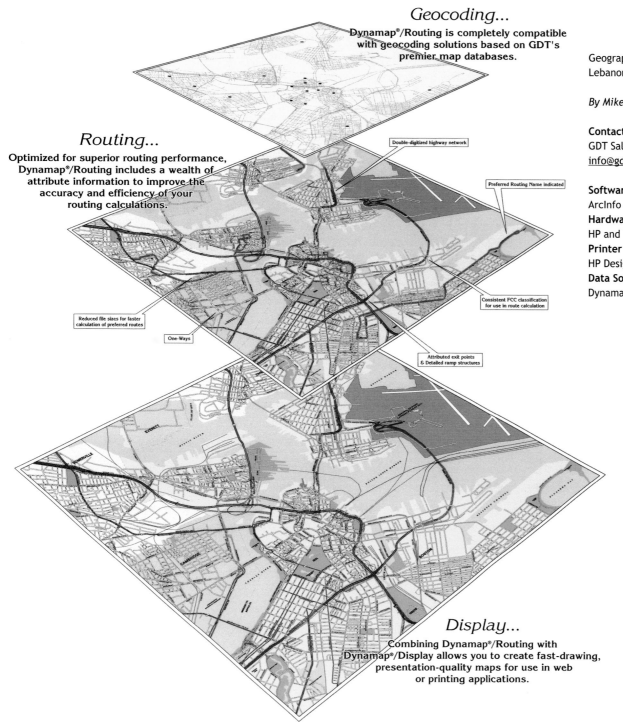

Geocoding...
Dynamap®/Routing is completely compatible with geocoding solutions based on GDT's premier map databases.

Routing...
Optimized for superior routing performance, Dynamap®/Routing includes a wealth of attribute information to improve the accuracy and efficiency of your routing calculations.

Double-digitized highway network

Preferred Routing Name indicated

Reduced file sizes for faster calculation of preferred routes

One-Ways

Consistent FCC classification for use in route calculation

Attributed exit points & Detailed ramp structures

Display...
Combining Dynamap®/Routing with Dynamap®/Display allows you to create fast-drawing, presentation-quality maps for use in web or printing applications.

Geographic Data Technology, Inc.
Lebanon, New Hampshire

By Mike Griffen

Contact
GDT Sales
info@gdt1.com

Software
ArcInfo
Hardware
HP and Windows NT
Printer
HP DesignJet
Data Source(s)
Dynamap/Routing

7 | business

This map displays the different elements important to a routing application and demonstrates the application of Dynamap/Routing as a modular component for routing solutions. Dynamap/Routing includes elements such as turn restrictions, one-way indicators, detailed ramping structures, and multilane highway representation, all of which are crucial for routing applications.

METRO (Method for Elimination of Tilt and Relief Displacement in Orthophotography)

Analytical Surveys, Inc.
Colorado Springs, Colorado

By Brad Barnell

Contact
Brad Barnell
bbarnell@anlt.com

Software
ArcView GIS and ArcView 3D Analyst™
Hardware
PC
Printer
DisplayMaker 6000
Data Source(s)
Photogrammetry

Analytical Surveys, Inc. (ASI), created this three-dimensional building and terrain model of downtown San Diego for the San Diego Data Processing Corporation (SDDPC). It was part of a project to update their digital terrain model (DTM) data and produce new digital orthophoto imagery for a 337-square-mile area within the city of San Diego.

The radial displacement of the tall buildings in the orthophoto imagery has been eliminated using an innovative production process developed by ASI called METRO. This automated process rectifies the position of the buildings in their true location as defined by the three-dimensional building and terrain model. A sophisticated mosaicking process combines imagery from the adjacent aerial photographs of the area to fill in the "blind spots" on the ground surrounding the buildings and/or bridges. The resulting seamless orthophoto image contains buildings without lean and sidewalks and roads that are completely visible.

The main image was created by draping the orthophoto imagery over the building and terrain model using ArcView 3D Analyst. Users may "fly around" the downtown San Diego area and view ground features on all four building sides.

Mojave Desert Ecosystem Shaded Relief Map

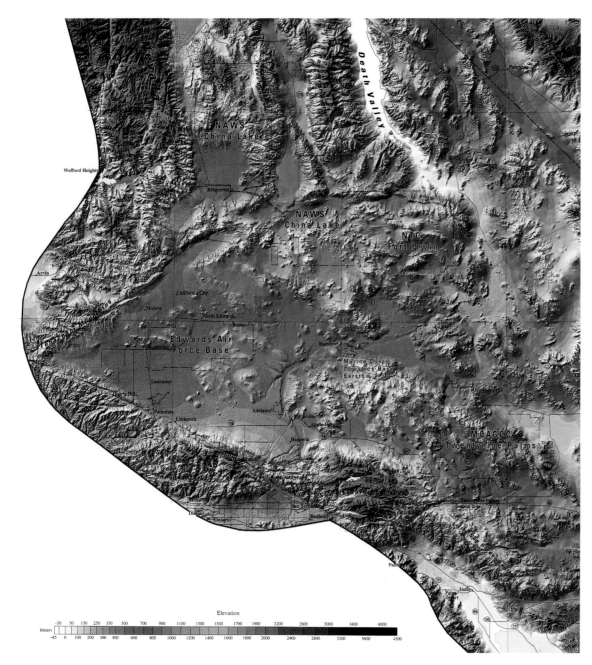

U.S. Department of Defense
 Mojave Desert Ecosystem Program
Barstow, California

By Robert Johnson and Dr. R. Douglas Ramsey

Contact
Clarence Everly
everlyc@mojavedata.gov

Software
ArcInfo 7.2.1, ARC GRID™, and Adobe Photoshop 5.0
Hardware
Sun workstation and Apple PowerMac
Printer
Offset-printed
Data Source(s)
U.S. Geological Survey; U.S. Department of Interior, Bureau of Land Management; and project boundary based on the Mojave Desert Section

The shaded relief map database was prepared from U.S. Geological Survey digital elevation model data and consists of digital terrain elevation data in a digital raster form. The parameters are solar elevation—25 degrees, azimuth—315 degrees, exaggeration—five times, and ambient light—0.5.

A color roam-around version of this map is included on the elevation CD–ROM of the Mojave Desert Ecosystem Program CD–ROM set in the elevation directory under the file name "mojsr125.tif." A three-arc-second, 93.218-meter version is on the introductory CD–ROM in the elevation directory under the file name "mojrel."

A Perspective View of the Pacific Northwest

U.S. Department of the Interior
 Bureau of Land Management
Portland, Oregon

By Jeffery Nighbert

Contact
Jeffery Nighbert
jnighbert@or.blm.gov

Software
ArcInfo 8, ARCPLOT™, and ARC GRID
Hardware
Dell Windows NT workstation
Printer
ENCAD NovaJet 60
Data Source(s)
U.S. Geological Survey 1-kilometer
and 90-meter digital elevation
models of North America

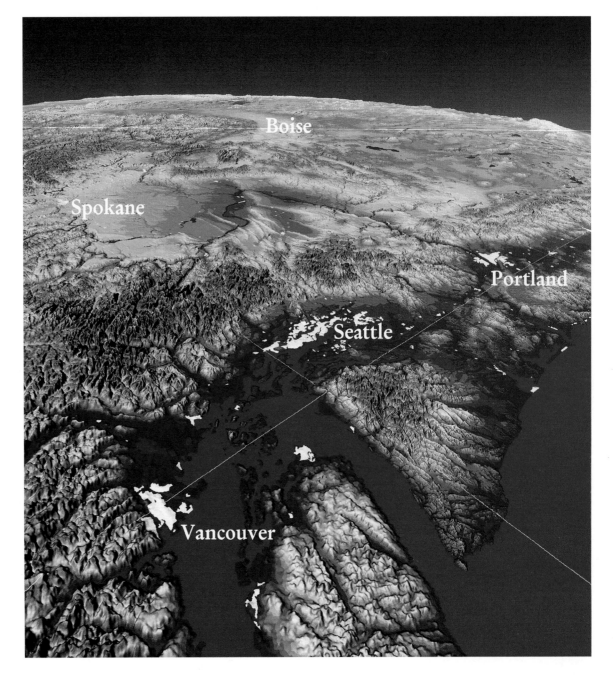

Perspective views offer map users the opportunity to see the world in new and exciting ways that are not available from traditional maps. Each view illuminates different geographical relationships and promotes discovery and understanding of the world around us. Often we base our perception and interpretation of the world around us on the traditional map. Computer GIS mapping takes us a step beyond traditional maps to see new horizons. The map authors created this map to encourage use of the Surface Display and Analysis commands in the ArcInfo GIS package.

The maps illustrate how the Columbia basin would look from 500,000 meters in the air above the western Canada coast. An accompanying explanation plate illustrates perspective viewing from the eight cardinal directions.

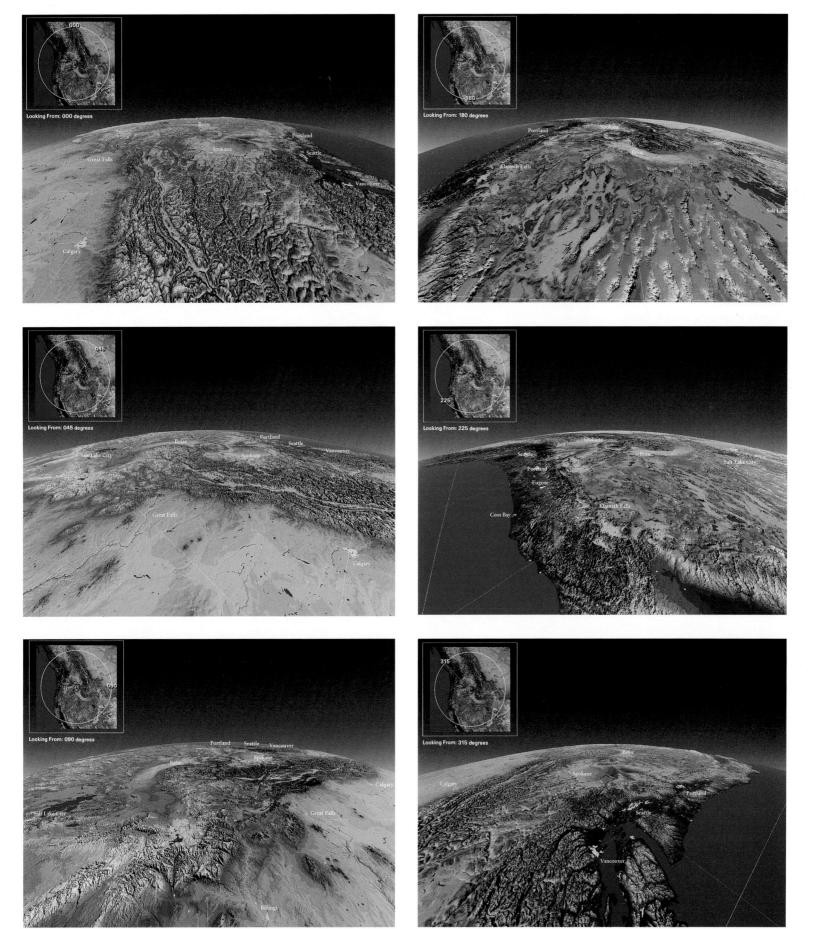

GEOCR500—Set of Digital Geoscientific Maps of the Czech Republic, 1:500,000

ARCDATA PRAHA
Prague, Czech Republic

By Jan Vodnansky

Contact
Jan Vodnansky
vodnansky@arcdata.cz

Software
ArcView GIS 3, ArcView Spatial
Analyst 1.1, and ArcPress™ 2
Hardware
DEC PW Pentium II
Printer
HP DesignJet 750C+
Data Source(s)
Czech Geological Survey, Geofyzika,
GISAT, Land Survey Office, and
ARCDATA PRAHA

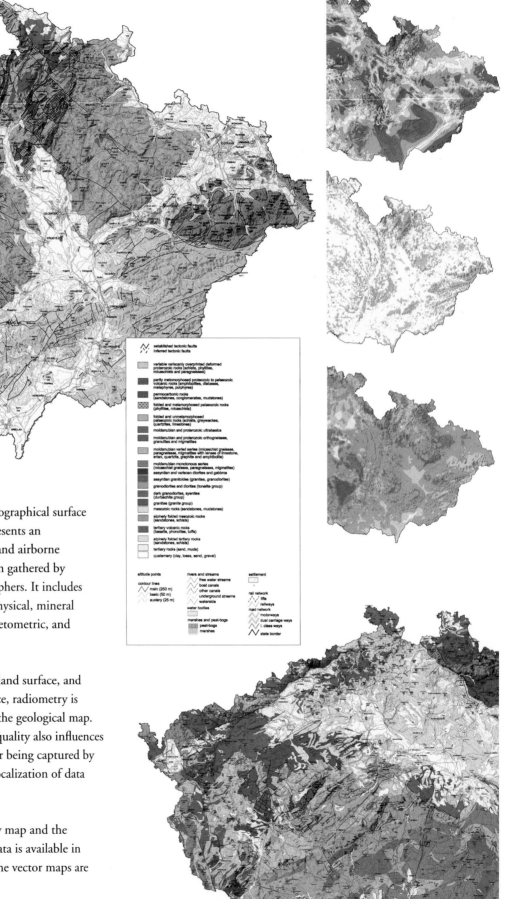

GEOCR500, which is available on CD, covers the geographical surface
of the entire territory of the Czech Republic and represents an
array of measurements collected by ground, satellite, and airborne
surveys. The data is a unique collection of information gathered by
geologists, geophysicists, land surveyors, and cartographers. It includes
topographical, satellite, land surface, geological, geophysical, mineral
waters, metallogenetic, radiometric, radon risk, magnetometric, and
gravity maps.

Each of the maps uses specific methods to depict the land surface, and
the data of individual maps is interwoven. For instance, radiometry is
influenced by the rock quality, which is indicated on the geological map.
Radiometry is closely related to radon risk. The rock quality also influences
gravimetry, magnetometry, and morphology, the latter being captured by
topographical and satellite maps, which enables the localization of data
on other maps.

All of the maps are in vector format except the gravity map and the
satellite map, which are depicted by symbology. All data is available in
national coordinate systems (S-JTSK and S42), and the vector maps are
available in geographic coordinates.

General Maps 1:300,000

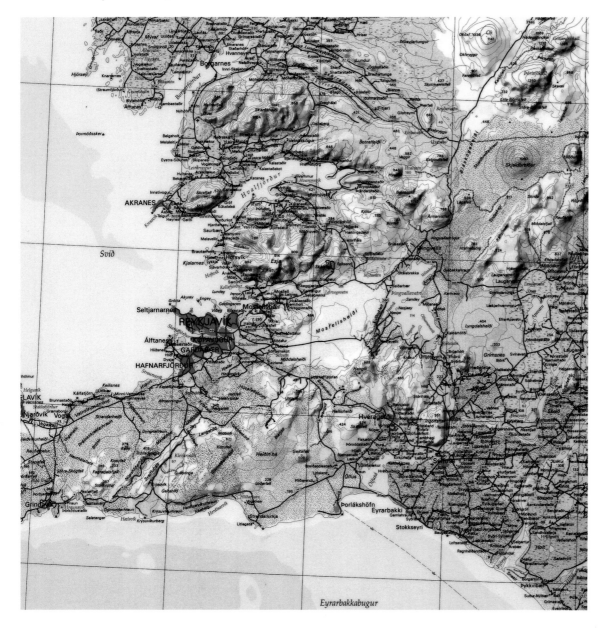

Mál og menning
Reykjavík, Iceland

By Hans Hansen

Contact
Hans Hansen
hans@ni.is

Software
ArcInfo 7.1.2
Hardware
PC
Printer
Professionally printed
Data Source(s)
Icelandic Institute of Natural History,
National Land Survey of Iceland,
Science Institute, and University of
Iceland

These four-part general maps in 1:300,000 scale are an innovative set, dividing Iceland into four equal parts. All the maps overlap and include up-to-date information on the road network, campsites, swimming pools, and museums. They were the recipient of the first place award for Best Cartographic Design, Map Series or Atlas at ESRI's 1999 International User Conference.

On the back of the maps, there are details in four languages, color photographs of Iceland's most famous attractions, and a table of road distances. The maps are sold in a protective plastic wallet. As printed versions of the database, these maps are easily updated and reprinted every year.

Selected Hydrologic Features of Lake Tahoe Basin and Surrounding Area, California and Nevada, 1998

U.S. Geological Survey
Carson City, Nevada

By *John E. Hughes Clarke,*
Peter Dartnell, James Gardner,
Larry A. Mayer, Timothy G. Rowe,
J. LaRue Smith, and J. Christopher
Stone

Contact
Public Information Assistant
gs-w-nvpublic-info@usgs.gov

Software
ArcInfo 7.1.1 and Adobe Illustrator
Version 8.0
Hardware
SGI and Macintosh
Printer
Varied
Data Source(s)
U.S. Geological Survey, National
Oceanic and Atmospheric
Administration, and U.S. Forest
Service

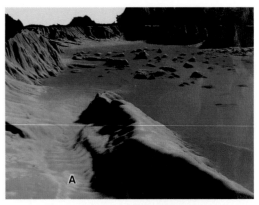

The last study of the bottom of Lake Tahoe, which straddles the California/Nevada border in the Sierra Nevada mountain range, was conducted in 1923. Although that work yielded good information about the lake's depths, there was not much detail about sedimentation processes and geological formations.

At the Lake Tahoe Presidential Forum in 1997, the U.S. Geological Survey (USGS) made a commitment to develop a new bathymetry data set, and in August 1998, a USGS team began a new mapping project of the lake. This effort was based on state-of-the-art technology and methods. The team used a high-resolution multibeam sonar system in swaths over the lake bottom as they navigated a boat back and forth. At the same time, a global positioning system fed positioning information into a database. They collected more than 65 million data points.

The bathymetry data collected during the team's 12-day effort is shown on these maps. They illustrate unique features and depths in Lake Tahoe's bottom, which is depicted within a shaded relief map of the surrounding basin.

This data will enable further research to help provide a better understanding of the geologic and hydrologic processes within the Lake Tahoe basin. The digital data used to prepare this map and images of the bottom of Lake Tahoe is available online at http://tahoe.usgs.gov.

Highlighting Physical Characteristics Using Shaded Relief

Utah School and Institutional Trust
 Lands Administration
Salt Lake City, Utah

By Jeff Roe

Contact
Jeff Roe
tlajar@lands5.state.ut.us

Software
ArcInfo, ARC GRID, ArcPress, and
ARC TIN™
Hardware
HP 9000/C100 workstation
Printer
HP DesignJet 2500CP
Data Source(s)
U.S. Geological Survey GeoData Site
and ESRI Atlas GIS™

15 | cartography

The Utah School and Institutional Trust Lands Administration manages 3.2 million acres of surface lands and 4.7 million acres of mineral lands for its beneficiaries, of which more than 95 percent goes to the school fund. Those lands are scattered all over the state covering a variety of geographic features. Since Utah varies so much topographically, the GIS staff realized that incorporating physical relief into map production would aid in visualizing the topographic nature of the lands it manages. They added shaded relief as a GIS technique.

Shaded relief can be applied to any color theme. Color themes come from polygridding coverages or creating color elevation bands with digital elevation model information. Color themes can be manipulated and combined in many ways to create the final color image. The shaded relief element added to any color theme conveys a quicker image of the topology than contour lines alone.

Perspective views also help visualize project areas and bring understanding to the physical nature of the lands. The agency works with 90-meter (3-arc-second), 30-meter, and 10-meter digital elevation model data depending on the resolution and extent of the area. These techniques are still under development at the Trust Lands Administration, and more useful and meaningful products will continue to emerge.

The Panama Canal

GeoInfo Internacional
Panama City, Panama

By Ermelinda Munoz

Contact
Ermelinda Munoz
emunoz@pananet.com

Software
ArcInfo 7.2, ARC GRID, ARC TIN,
ArcView GIS 3.1, and ArcView Spatial
Analyst
Hardware
Intel Pentium II workstation
Printer
HP DesignJet 650C
Data Source(s)
Instituto Geografico Tommy Guardia,
GeoInfo SA, Smithsonian Tropical
Research Institute, and Panama
Canal Commission

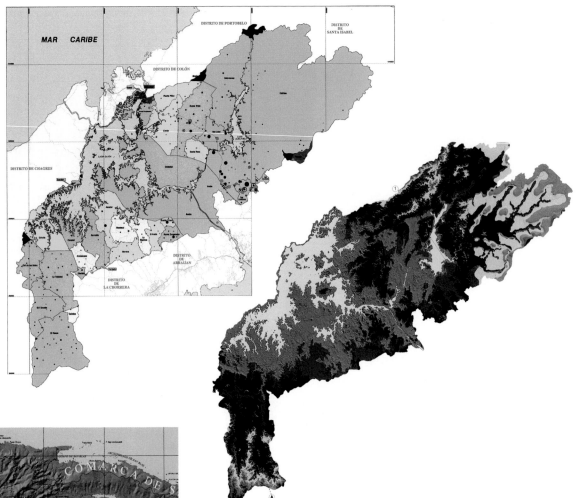

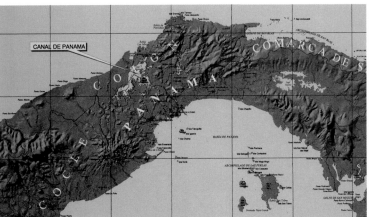

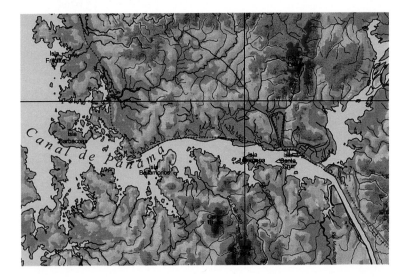

This set of maps was produced to illustrate the local characteristics such as population, access roads, elevations, locks, lakes, rivers, and the watershed of the Panama Canal.

The canal is 50 miles long from the Atlantic to the Pacific oceans. It is located on one of the narrowest and lowest saddles of the long, mountainous isthmus that joins the continents of North and South America. It takes eight to 10 hours for a ship to transit the canal.

Among the principal features of the canal is Gatun Lake, a central man-made lake stretching nearly all the way across the isthmus. The Gallard Cut is an 8.5-mile excavation through the Continental Divide, which connects Gatun Lake to the Pedro Miguel Locks. The Gatun Locks on the Atlantic and the Pedro Miguel and Miraflores Locks on the Pacific raise ships from sea level to the level of Gatun Lake.

When the canal was built in 1914, the Gatun Dam was the largest earth dam ever constructed, and Gatun Lake was the largest man-made lake in the world. The three sets of locks were the most massive concrete structures ever built.

Database-Driven Cartography

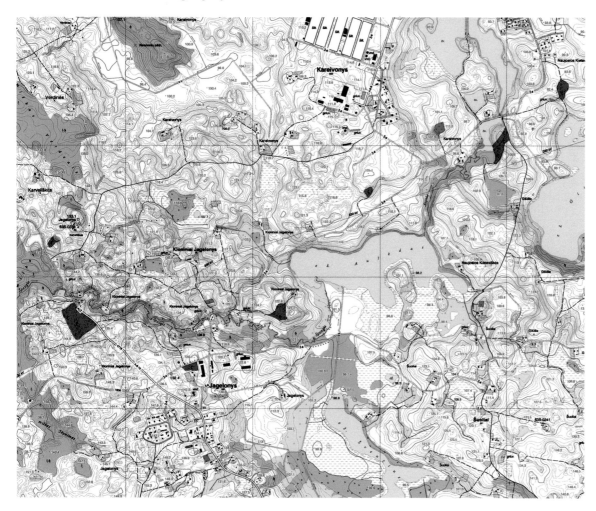

Institute of Aerial Geodesy
Kaunas, Lithuania

By Vilma Davalgaite and Vilius Zilevicius

Contact
Vilma Davalgaite
agi_vd@kaunas.omnitel.net

Software
ArcInfo 7.2.1 and TopoMap10
Hardware
PC, Windows NT Version 4.0
Printer
HP DesignJet 2000CP
Data Source(s)
Data collected from different processes and sources

The main goal of the mapping process was creating a cartographic database (CDB) instead of producing map sheets. Three years ago, the Institute of Aerial Geodesy in Kaunas, Lithuania, introduced a database-driven cartography process at a national scale. The main source for vector data extraction is orthophotography. In some more complex areas, stereodigitizing is in use.

Data is collected for the CDB, which is GIS oriented. Editing is done on a Windows NT ArcInfo workstation, and data in editing is managed on a UNIX ArcInfo workstation. Map sheets are generated directly from the CDB using the TopoMap10 program in an ArcInfo environment. Members of the Institute of Aerial Geodesy and HNIT– Baltic GeoInfoServisas in Vilnius, Lithuania, developed TopoMap10.

Recent Cartography Work by MapQuest.com

MapQuest.com, Inc.
Mountville, Pennsylvania

*By Michael Dangermond, J. Fix,
D. Straub, and K. Winters*

Contact
Michael Dangermond
mdangermond@mapquest.com

Software
ArcInfo 7.2 and Adobe Illustrator
Hardware
Windows NT and Macintosh
Printer
HP DesignJet 2500CP
Data Source(s)
Digital Chart of the World, GTOPO30,
and MapQuest North America
database

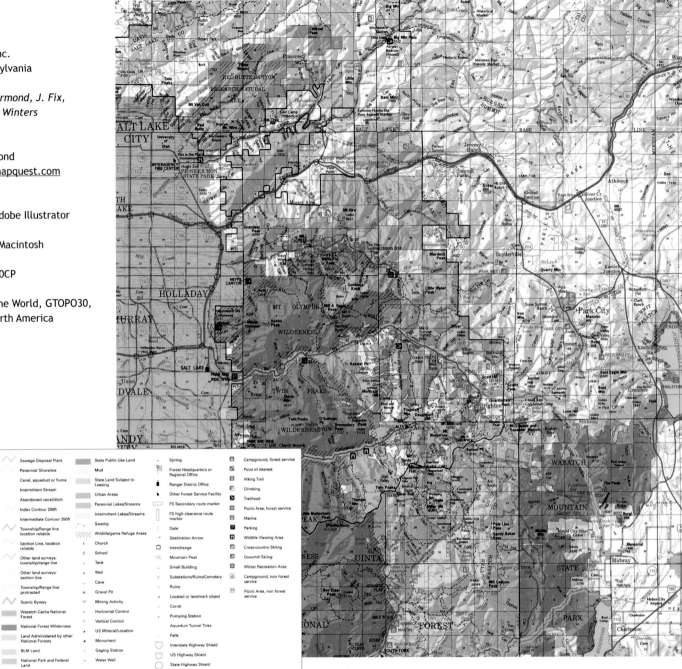

MapQuest GIS engineering services support the MapQuest.com cartographers, artists, and software engineers with quality databases and innovative methods of data processing for exceptional mapmaking. Exhibited on these pages are some of the interesting projects that the MapQuest.com GIS group has supported.

MapQuest.com has put a great deal of energy into leveraging ArcInfo as a support tool for cartography including improving the quality and flexibility of shaded-relief products from digital sources. In addition, the MapQuest.com GIS group uses ArcInfo to continuously update its cartographic databases. The North American database (NADB) of United States, Canadian, and Mexican highways and points of interest is used for both paper-media cartography and its digital mapping Web site.

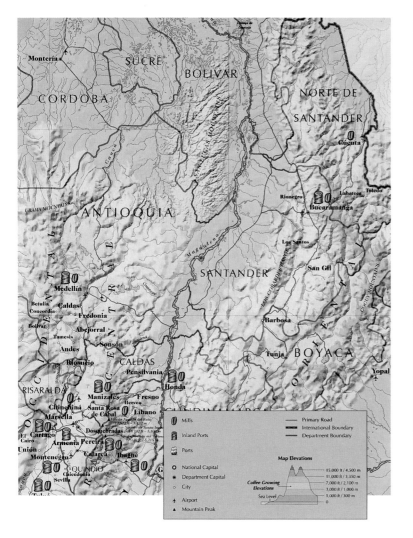

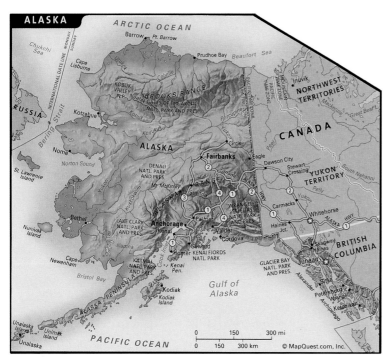

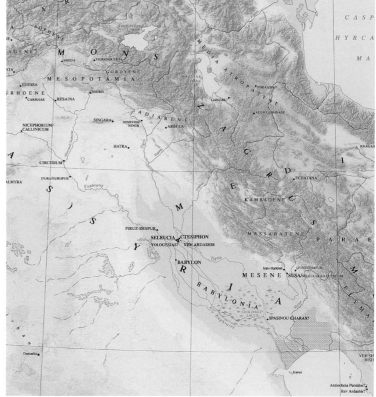

Geopolitical Map of Greece

Terra Ltd.
Athens, Greece

By Thanos Doganis, Apostolos Marnieros, and Yannis Roukoutakis

Contact
Thanos Doganis
thanos@terra.gr

Software
ArcInfo 7.1.2
Hardware
DEC Alpha 600 workstation
Printer
HP DesignJet 2500CP
Data Source(s)
Topographic maps at 1:500,000 scale and 1:250,000 scale, bathymetry at 1:1,000,000 scale, and Terra's digital data set

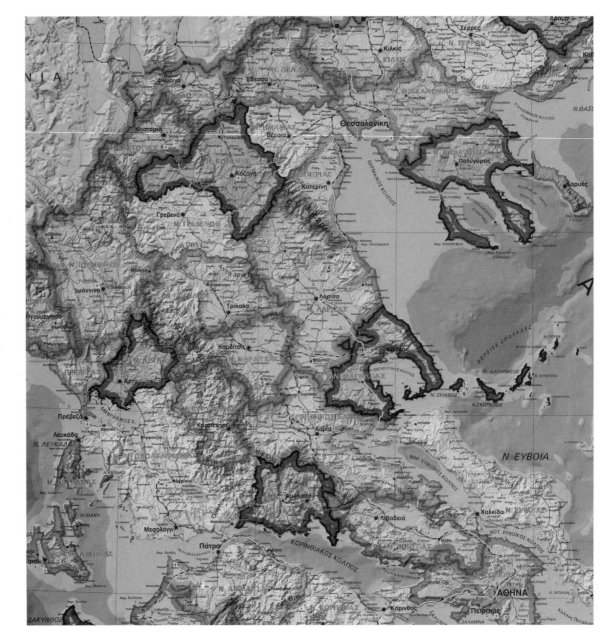

This map was designed and targeted for students of Hellenic Schools. The goal was to combine administrative information (such as boundaries of prefectures—nomoi in Greek) and the relief of the country. This was done to enable pupils to distinguish between flat and mountainous areas. This information, combined with details about each area's economy and culture, will help to give students a more comprehensive understanding of the land.

The morphology of the seafloor is detailed with reefs, canyons, and mountains that continue below sea level. The Hellenic arc and other major faults, which cause the earthquakes in this area, are visible on the map.

World Maps

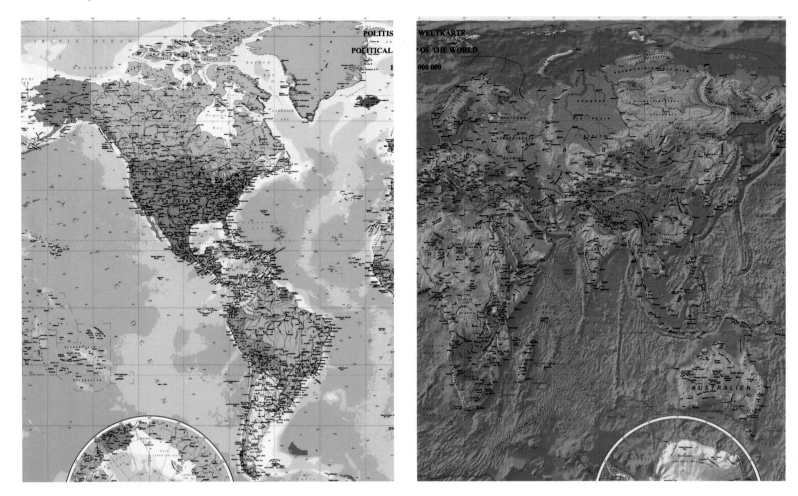

Political Map of the World—Existing geographic data was used to prepare this digital map of the world. The world is shown in Gall cylindrical projection and the polar areas are displayed in Postel azimuthal projection. Data from ESRI's Digital Chart of the World was merged to thematic coverages and generalized. The map contains countries distinguished by colors, boundaries, cities, airports, roads, railways, polar scientific bases, elevation points, rivers, and bodies of water.

Different signatures were used to label cities according to population and to denote state capitals. Feature annotations were generated from the database. Country annotations are in German and other annotations are in English. The legend is available in German, English, and Czech. The depth division in ocean areas is represented by bathymetry prepared using digital elevation model data from the U.S. Geological Survey.

Physical Map of the World—The world is shown in Gall cylindrical projection and the polar areas are displayed in Postel azimuthal projection. The map displays mountain regions, lowlands, depressions, volcanoes, ground elevations, water areas, glaciers, sea ice-ups and ice drift, jurisdictional boundaries, and cities. There are annotations of islands, rivers, lakes, seas, oceans, glaciers, orographic complexes, volcanoes, and cities. All of the annotations are in English except for the names of islands and the orographic complexes, which are in German. Capital cities are differentiated by colors.

The division of the continental surface is illustrated by shaded hypsometry, and the seafloor is represented by shaded bathymetry. The surface of the earth was processed using digital elevation model data from the U.S. Geological Survey.

T-Mapy
Hradec Dralove, Czech Republic
SHOCart
Zlin, Czech Republic

By Jan Langr and Jan Potstejnsky

Contact
Jan Langr
jala@tmapy.cz

Software
ArcInfo 7.2.1, ArcView GIS 3.1, and ArcView Spatial Analyst
Hardware
Sun Ultra 5 workstation and Pentium II PC
Printer
HP DesignJet 2000CP
Data Source(s)
ESRI's Digital Chart of the World and U.S. Geological Survey

National Atlas of the United States: Shaded Relief of North America

U.S. Geological Survey and
  The National Atlas of the United
  States of America™
Sioux Falls, South Dakota

By Peter J. De Vincentis,
John A. Hutchinson, and
Dean J. Tyler

Contact
John A. Hutchinson
hutch@edcmail.cr.usgs.gov

Software
ArcInfo, ARC GRID, Adobe Illustrator,
and Adobe Photoshop
Hardware
Sun and SGI UNIX servers and Apple
Macintosh G3
Printer
Harris five-color offset press
Data Source(s)
GTOPO30 global elevation data set,
National Atlas data set, and Digital
Chart of the World

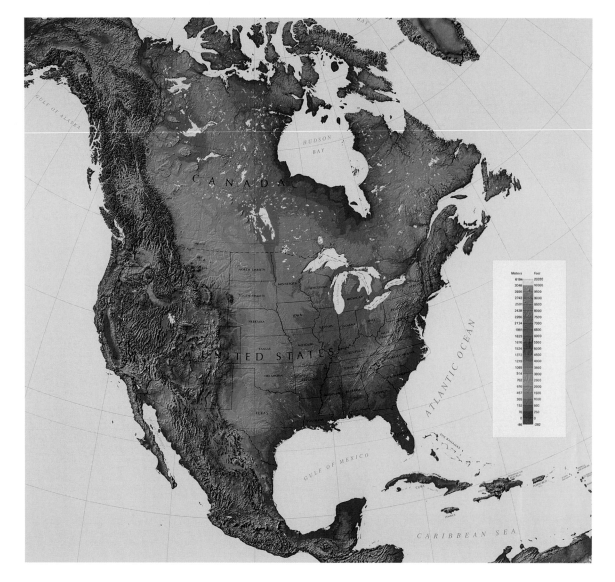

This shaded relief map of North America was created at 1:10,000,000 scale for the National Atlas of the United States (www.nationalatlas.gov). The elevation tints and relief shading were generated from GTOPO30, a global digital elevation data set covering the land surface of the entire earth.

The depiction of shaded relief was generated from GTOPO30 with the ArcHillshade command, using illumination from the northwest (azimuth 315 degrees) by a simulated sun 45 degrees above the horizon and a vertical exaggeration factor of 10x. Within each elevation range, the lightest color tones represent fully illuminated steep slopes and the darkest tones represent steep areas in shadow. Intermediate color tones show areas of relatively gentle topography.

Shorelines and boundaries are from the 1:2,000,000-scale National Atlas data set and the Digital Chart of the World (DCW). The lakes shown were selected from DCW and cross-checked in ARCEDIT against printed maps. A generalization technique based on the ArcBuffer command

was used to simplify all shorelines by removing small features; the ArcGeneralize command was then applied for final smoothing.

While all of the processing of the geographic data from this map was done within ArcInfo, the cartographic layout and composition of the map were completed in Adobe Photoshop and Illustrator. The shaded relief image was converted to a BSQ file with the ARC GRID command, loaded into Adobe Photoshop, and converted from red, green, and blue display colors to cyan, magenta, yellow, and black for printing. PostScript files were created in ARCPLOT from the shoreline and boundary data sets for processing into Adobe Illustrator. The final map layout and text placement were done in Illustrator.

Color separations for the map were contracted through the Government Printing Office. The map is for sale through the U.S. Geological Survey Branch of Information Services, Box 25286, Denver, Colorado 80225. The stock number is TUS5682.

Strategic Planning for Habitat Restoration and Protection

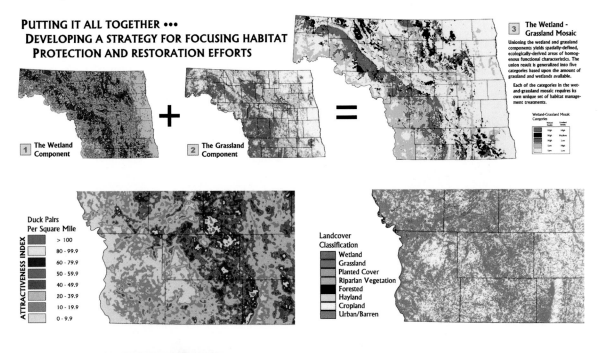

PUTTING IT ALL TOGETHER •••
DEVELOPING A STRATEGY FOR FOCUSING HABITAT PROTECTION AND RESTORATION EFFORTS

1 The Wetland Component **+** **2** The Grassland Component **=**

3 The Wetland - Grassland Mosaic

Unioning the wetland and grassland components yields spatially-defined, ecologically-derived areas of homogenous functional characteristics. The union result is generalized into five categories based upon the amount of grassland and wetlands available.

Each of the categories in the wet-and-grassland mosaic requires its own unique set of habitat management treatments.

Wetland-Grassland Mosaic Categories

Ducks Unlimited, Inc.
Bismarck, North Dakota

By Darin Blunck, Chuck Loesch, and Duane Pool

Contact
Duane Pool
dpool@ducks.org

Software
ArcInfo 7, ArcView GIS 3.1, ArcView Spatial Analyst 1.1, and ERDAS IMAGINE Version 8.3
Hardware
Sun Ultra 1 workstation, Dell Windows NT workstation, and Dell Latitude
Data Source(s)
Landsat Thematic Mapper (TM); U.S. Fish and Wildlife Service (USFWS), National Wetlands Inventory; U.S. Geological Survey (USGS), Northern Prairie Wildlife Research Center; Suremap; ESRI, USGS, and USFWS Habitat and Population Evaluation Team (HAPET)

Duck Pairs Per Square Mile

ATTRACTIVENESS INDEX
- > 100
- 80 - 99.9
- 60 - 79.9
- 50 - 59.9
- 40 - 49.9
- 20 - 39.9
- 10 - 19.9
- 0 - 9.9

Landcover Classification
- Wetland
- Grassland
- Planted Cover
- Riparian Vegetation
- Forested
- Hayland
- Cropland
- Urban/Barren

Perennial Cover Percent per 4 mi²
- > 75 %
- 50 - 74 %
- 40 - 49 %
- 30 - 39 %
- 20 - 29 %
- 10 - 19 %
- 0 - 9 %

The Grasslands for Tomorrow initiative area focuses on protecting and restoring the prairies of the Central and Great Plains. This area encompasses the most important waterfowl habitat on the continent—the prairie pothole region—where retreating glaciers left a landscape dotted with millions of smallish wetlands critical to breeding ducks.

Yet wetlands provide only half of the ecological needs of breeding ducks. Ducks also require nearby blocks of grassland in which to nest. In much of the plains region, grasslands have been converted to row crops or overutilized by intensive grazing. The paucity of quality nesting cover has forced hens to nest in less ideal areas where they and their eggs are easy targets for predators.

Ducks Unlimited, through its habitat conservation programs in the Grasslands for Tomorrow initiative area, is at work restoring and protecting landscapes for the benefit of North America's waterfowl and other prairie species.

The land protection program is one example of where Ducks Unlimited has integrated GIS into all levels of its conservation program—from outlining conceptual program objectives to narrowing the focus at the subcounty level when locating potential projects.

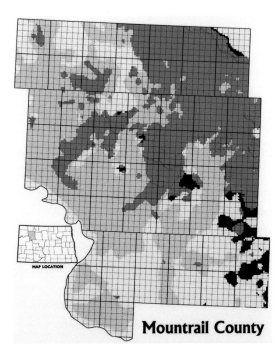

MAP LOCATION

Mountrail County

Biodiversity Conservation Planning in the Central Appalachian Mountains

The Nature Conservancy–Eastern
 Conservation Science
Boston, Massachusetts

*By Frank Biasi, Mike Merrill, and
Arlene Olivero*

Contact
Arlene Olivero
arlene_olivero@tnc.org

Software
ArcInfo 7.2.1 and ArcView GIS 3.1
Hardware
Dell Dimension XPS r400
Printer
HP DesignJet 650C
Data Source(s)
Macon USA TIGER 1994; U.S.
Geological Survey (USGS) digital
elevation models; Multi-Resolution
Land Characteristics land cover data
set; federal, state, and private
conservation data; The Nature
Conservancy-Eastern Conservation
Science; Natural Heritage Program;
USGS digital line graphs; USGS
Geographic Names Information
System database 1998; and ESRI
ArcData^SM 1998

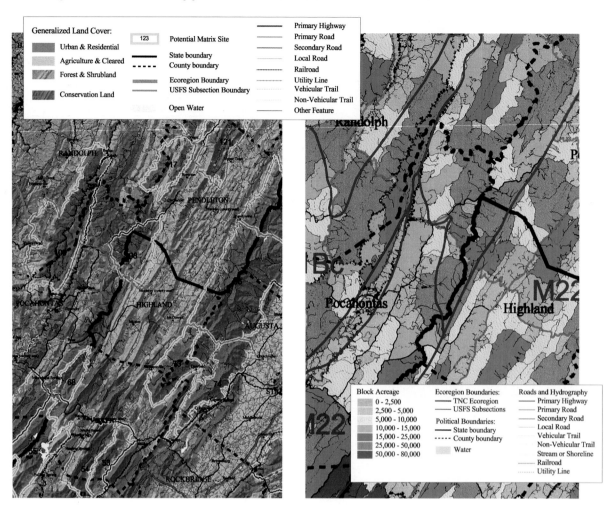

Central Appalachian Ecoregion, Functional Landscapes—Functional landscapes are large contiguous areas with ample size and natural conditions to facilitate the maintenance of ecological processes and viable occurrences of matrix forest communities. They also have embedded large- and small-patch communities and embedded species populations. The core functional landscape areas on this map, matrix sites, are embedded within a network of other conservation and natural lands that together form a semicontiguous functional network. This network is presumed to support abiotic and biotic patterns and processes across their natural range of variability.

The core matrix sites were defined by an ecoregional planning process, which used GIS analysis and expert interviews to select and rank large areas of natural land. The minimum size for a matrix site was based on the area needed to absorb and recover from the characteristic large, infrequent disturbances that occur in this ecoregion. The size criterion was intended to maximize the probability that sites contained a full complement and multiple breeding populations of all associated fauna and flora, particularly species that require interior forest conditions.

The matrix site selection process followed four sequential steps: (1) develop a set of all potential matrix sites based on a GIS analysis of road-bounded "blocks" greater than 15,000 acres; (2) determine which blocks qualify for inclusion by assessing the condition of each potential block through GIS and expert analysis; (3) assess the composition of ecological land units (ELUs) within each block to cluster the blocks into ecologically similar groups based on ELU composition; and (4) prioritize blocks within each ELU group into Tier 1 or Tier 2 conservation priority based on diversity, condition, proximity to other features, feasibility, and threat.

A range of ecological attributes describing the size, condition, diversity, and landscape context of each block were collected and used in conservation planning and site selection. A size index was created based on the total block area as well as the area of core and edge of the block derived by buffering roads.

Once attributed, the ecoblocks can be used in automated and interactive conservation site selection processes. The ecoblock approach enables entire landscapes to be assessed and compared in a standard scientific manner. Varying weights can be applied to each index or component attribute to compare and select sites based on alternate conservation goals. Ecoblocks are also useful in targeting areas for field and remote sensing inventories.

Bay Area EcoAtlas Past and Present

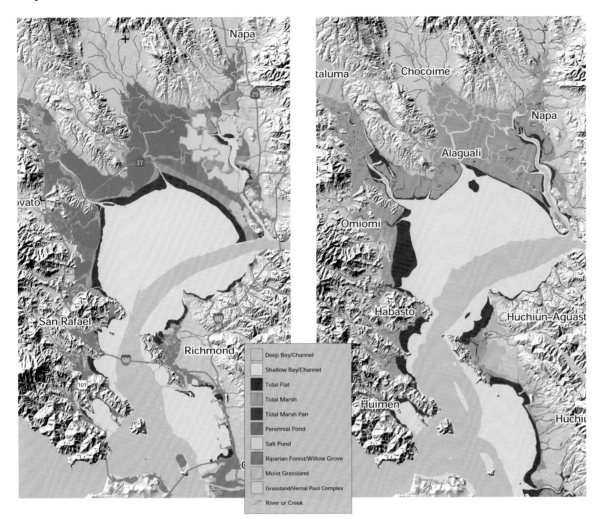

Legend:
- Deep Bay/Channel
- Shallow Bay/Channel
- Tidal Flat
- Tidal Marsh
- Tidal Marsh Pan
- Perennial Pond
- Salt Pond
- Riparian Forest/Willow Grove
- Moist Grassland
- Grassland/Vernal Pool Complex
- River or Creek

San Francisco Estuary Institute
Richmond, California

*By Elise Brewster, Josh Collins,
Zoltan Der, and Robin Grossinger*

Contact
Zoltan Der
zoltan@sfei.org

Software
ArcInfo 7.2.1, ArcView GIS 3.1, Adobe
Illustrator Version 7.01, MapPublisher
Version 3.0, ImageAlchemy Version 1.11,
and PhotoShop Version 1.1
Hardware
Dell 410 Windows NT workstation and
Alpha Digital UNIX
Printer
ENCAD Pro
Data Source(s)
U.S. Coast Survey, U.S. Geological
Survey, U.S. Department of
Agriculture, Spanish diseños,
explorer's journals, local archives,
Randall Milliken, California State
Lands Commission, U.S. Fish and
Wildlife Service, U.S. National
Aeronautical and Space
Administration, and local experts

These two maps represent past and present views of the San Francisco Bay region. The past view describes the bays, baylands, and adjacent habitats, as they appeared about 200 years ago, when Europeans first arrived in the region. The present view describes the same features, circa 1998.

Each major tributary to the bay area had tidal flats and tidal marshes arrayed along a salinity gradient created by local runoff. These subregional and local gradients of salinity created a complex system of tributary estuaries arrayed along the major salinity gradient between the Golden Gate and the delta, which supported great physical and biological diversity. Each day, as the tide went out, tidal flats emerged along the margins of the bays and larger tidal channels.

Native Americans lived near the estuary for thousands of years. About 200 years ago, people began to alter the bay area landscape in major ways. Europeans first sighted San Francisco Bay in 1769, and until 1821, the missionaries used the lands around the estuary for grazing cattle and sheep. The first large-scale changes in the region's natural habitats are associated with this land use. They included the clearing of oak woodlands, the conversion of large areas of native perennial grasslands to pastures of nonnative invasive annual grasses, and the advent of excessive erosion from local hillsides and creek banks.

Beginning in the mid-1800s, large areas of the estuary's tidal marshes and mudflats were filled, diked, or drained to provide land for transportation and industrial development. By the early 1900s, grazing in Suisun had given way to more lucrative land uses, and farmers were producing a variety of crops along with livestock and dairy products. Eventually, as increasing salinity and, to a lesser extent, land subsidence made it difficult to regulate groundwater levels and soil salinity, agriculture began to fail. Today, the only farming remaining in the Suisun baylands is the production of oat hay on some 1,500 acres.

In South Bay, the baylands were never extensively diked for agriculture. Instead, large areas were reclaimed for salt production. Farmers began to produce crops in the moist grasslands adjacent to South Bay in the 1850s. But, as the human population of the subregion increased, most of the agricultural areas adjacent to the baylands were developed for residential and industrial uses. By the 1950s, there were only about one quarter of the historical amount of tidal marshes in the estuary. The loss of tidal marshes has continued, but at a much slower rate.

Supporting Nonprofits with GIS

GreenInfo Network
San Francisco, California

*By Brian Cohen, Aubrey Dugger,
Ezra Freeman, and Larry Orman*

Contact
Larry Orman
larry@greeninfo.org

Software
ArcView GIS 3.1, ArcView Spatial
Analyst, and ArcPress
Hardware
HP PC workstations
Printer
HP DesignJet 755CM and 2500CP

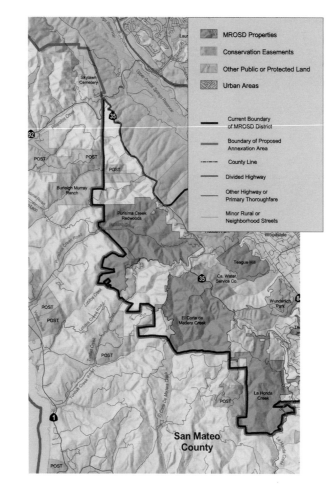

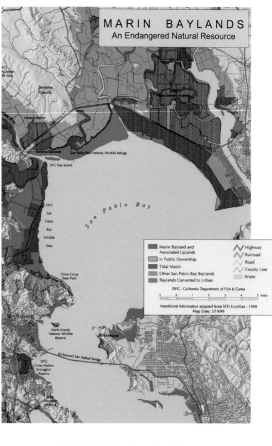

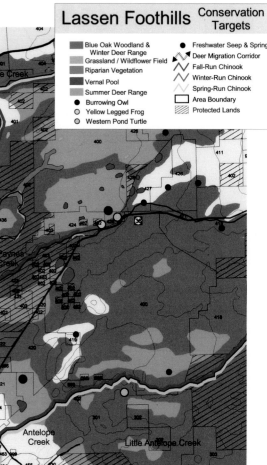

GreenInfo Network is a nonprofit organization providing mapping and information services to public interest groups. Based in San Francisco, GreenInfo Network supports environmental, social service, public health, and other client groups and governmental agencies, primarily in California.

The maps shown here represent a range of products created for GreenInfo's environmental client groups. For most of its projects, the focus is on making the maps cartographically effective and technically sound. This coincides with the objectives of many public interest groups seeking to change governmental policies or communicate more effectively with the public.

Lands of the San Mateo Coast, San Mateo County, California—This map was produced for the Midpeninsula Regional Open Space District to support a study of annexation into that district. Delicate shaded relief enables the overlay of primary thematic information in saturated color, and streams are symbolized with a line weight that gradually increases as they descend the watershed.

Marin Baylands, An Endangered Natural Resource, Marin and Sonoma Counties, California—Produced for the Marin Baylands Advocates, this map provides a vision of the issues facing wetlands in the North Bay area. The tabloid-sized map focuses attention on "at risk" lands and provides rich detail for locating small parcels.

Lassen Foothills, Conservation Targets, Lassen County, California—The California office of The Nature Conservancy uses this map, which depicts sensitive habitats, terrestrial and aquatic species, and wildlife movement patterns. Protected lands and private ownership information is also included.

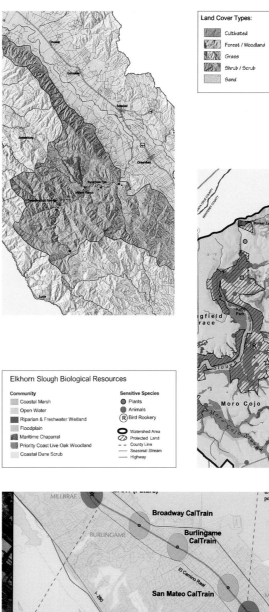

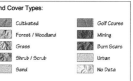

Land Cover Types:

	Cultivated		Golf Course
	Forest / Woodland		Mining
	Grass		Burn Scars
	Shrub / Scrub		Urban
	Sand		No Data

Elkhorn Slough Biological Resources

Community
- Coastal Marsh
- Open Water
- Riparian & Freshwater Wetland
- Floodplain
- Maritime Chaparral
- Priority Coast Live Oak Woodland
- Coastal Dune Scrub

Sensitive Species
- ● Plants
- ● Animals
- ® Bird Rookery

- ◯ Watershed Area
- ⬭ Protected Land
- — County Line
- —— Seasonal Stream
- —— Highway

Protected Public Land

Private Conservation Easements

Other Private Land

The San Francisco Bay Area map demonstrates the use of shaded relief in depicting land uses in the region. The map shows open space and other park lands currently in public ownership or under easement by private land trusts.

Location of Water Monitoring Stations, Santa Cruz and Monterey Counties, California—This map helps the Coastal Watershed Council, a nonprofit organization, to understand the resources and conservation planning issues in their area. The map was developed through detailed overlay analysis of vegetation and land use, combined with the location of water quality monitoring stations.

Elkhorn Slough Biological Resources, Monterey and San Benito Counties, California—Part of a comprehensive plan for managing the 43,000-acre Elkhorn Slough, the map identifies areas where particular habitats exist in this highly diverse and rich biotic landscape.

Potential for Transit Oriented Development—Produced for the Association of Bay Area Governments, this map is part of a reconnaissance study of options for building housing and other development near rail and ferry stations and along major bus corridors. The map shows areas within a quarter-mile radius (walking distance) of commuter train stations. Aerial photos of the buffer areas further illustrate potential sites.

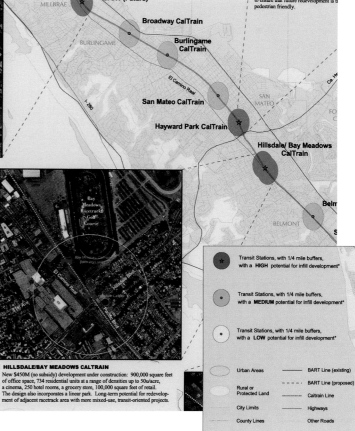

HILLSDALE/BAY MEADOWS CALTRAIN
New $450M (no subsidy) development under construction: 900,000 square feet of office space, 734 residential units at a range of densities up to 50u/acre, a cinema, 250 hotel rooms, a grocery store, 100,000 square feet of retail. The design also incorporates a linear park. Long-term potential for redevelopment of adjacent racetrack area with more mixed-use, transit-oriented projects.

- ★ Transit Stations, with 1/4 mile buffers, with a **HIGH** potential for infill development*
- ● Transit Stations, with 1/4 mile buffers, with a **MEDIUM** potential for infill development*
- • Transit Stations, with 1/4 mile buffers, with a **LOW** potential for infill development*

- Urban Areas — BART Line (existing)
- Rural or Protected Land — - - BART Line (proposed)
- City Limits — Caltrain Line
- County Lines — Highways
- Other Roads

Vietnam Herbicides and Wetlands: Locating the Hot Spots

International Crane Foundation
Baraboo, Wisconsin

*By Jeb Barzen, Mark Cheyne,
Le Duc Minh, Dorn Moore, and
Tran Triet*

Contact
Dorn Moore
dorn@savingcranes.org

Software
PC ARC/INFO® and ArcView GIS 3.1
Hardware
Trimble GeoExplorer
Printer
Commercial plotter

conservation

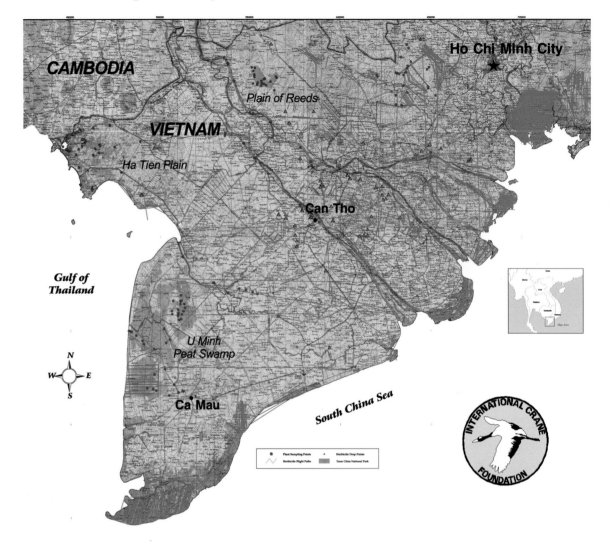

This map plots all herbicide missions flown in the Mekong Delta by the U.S. government during the Vietnam War. It also plots sampling sites from a recent study of freshwater wetland plants in the delta.

Herbicide data was acquired through the U.S. Freedom of Information Act to help service men and women determine if they were exposed to herbicides during their tours of duty. In order to plot the data, the flight lines and drop points were converted from the military coordinates to overlay on a scanned base map of Vietnam.

Tran Triet's goal in his Ph.D. work was to identify the factors that affect the distribution and abundance of freshwater wetland vegetation in the Mekong Delta. To do this, Triet sampled vegetation and environmental parameters in all the remnant wetlands that he could find in 1997 and 1998. The yellow circles on the map are the locations of all of Triet's sampling areas.

The herbicide map is relevant to Triet's study because herbicide applications had the potential to disrupt vegetation abundance or diversity. The map authors wanted to see how many vegetation sampling areas were affected by past herbicide missions. If Triet's study was to contribute to understanding freshwater wetland vegetation ecology of deltas throughout the world, they had to evaluate if herbicides had a significant influence.

U.S. military strategy becomes apparent from this data. The goal was to eliminate tree cover capable of hiding combatants. The easiest way to do this in the wetlands of the Mekong Delta was to drain the wetlands and burn them. Where drainage was not possible due to tidal flooding, vegetation was killed with herbicides. On this map, it is easy to recognize the areas significantly affected by herbicides.

Herbicides were also used intensively along canals in more populated areas of the delta. People tend to build homes on dredge spoils of canals, as these dikes are often the highest points of land. Trees and other thick foliage on the dredge spoils enabled boat traffic to be ambushed. Fire was ineffective, and herbicides were used to control the vegetation.

The ICF works worldwide to conserve cranes and the native ecosystems on which they depend. The organization is dedicated to providing experience, knowledge, and inspiration to people in their efforts to resolve threats to these ecosystems.

Madagascar Spiny Forest Ecoregion—Threat of Wood Stoves' Consumption in Main Cities

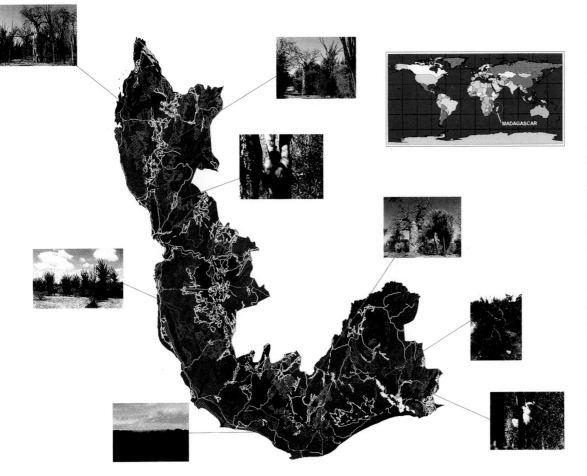

World Wildlife Fund
Antananarivo, Madagascar

By Jean Dominique Andriambahiny

Contact
Jean Dominique Andriambahiny
wwfrep@dts.mg

Software
ArcView GIS 3 and Idrisi
Hardware
PC Windows 95
Data Source(s)
National Institute of Census, National Institute of Geography and Hydrography, and National Ecological and Forestry Inventory

As part of the World Wildlife Fund (WWF) Living Planet 2000 Campaign, 234 ecoregions were identified worldwide (Global 200) for conservation initiatives. In 1997, the dry tropical and spiny forests found in eastern and southern Madagascar were determined to be among the priority ecoregions in Africa on which to focus present and future conservation investments.

Striking in appearance, the spiny forest includes the endemic family of *Didiereaceae,* which has four genera and 12 species. Numerous species are spiny and some reach heights of more than 10 meters. There are also several endemic species in the *Euphorbiaceae* family, which have interesting adaptations that enable them to survive in the harsh climate and poor soil conditions. Sifaka lemurs (*Lemur catta, Propithecus verreauxi verreauxi*), terrestrial tortoises (*Geocheleone radiata, Pyxis arachnoides*), and endemic reptiles and amphibians also inhabit the forest. Socioeconomic assessments identified various root causes for spiny forest degradation and loss. Urban energy was identified as a threat from within the ecoregion. In the eight urban areas of southern Madagascar, more than 90 percent of the fuel for cooking comes from charcoal and firewood collected in the spiny forest. The majority of the population in these urban areas depends on these cheap forest products for cooking because alternative cooking options (such as gas stoves) are prohibitively expensive. The peripheral forests of

these urban centers are under threat. The WWF in southern Madagascar developed a GIS-based model with ArcInfo, ArcView GIS, and Idrisi to assess where spiny forests still exist and to highlight the threats to their continued survival.

The model relies on the Madagascar census; National Institute of Geography and Hydrography (FTM) 1:500,000-scale topography, road, and track network data; National Ecological and Forestry Inventory (IEFN) 1:500,000-scale vegetation data; and local land use data. ArcView GIS was used to overlay the locations of spiny forest and sensitive zones. The resulting map highlights the most threatened habitat, with blue and dark blue where the spiny forest is more disturbed by firewood exploitation for urban consumption.

The WWF–Madagascar Dry Forest program hopes to add other resources to improve its assessments of implications of alternative development patterns in the landscape ecology of the dry forest ecoregion. A new Landsat image will be integrated to update the land use map, and the Defense Meteorological Satellite program (DMSP) Operational Linescan System (OLS) data from the U.S. Geological Survey will be integrated to analyze burning patterns and trends in southern Madagascar.

ESRI Sample Map Series

ESRI
Redlands, California

By Prashant Hedao

Contact
Prashant Hedao
phedao@esri.com

Software
ArcInfo 8
Hardware
Pentium running Windows NT
Printer
HP DesignJet 1055CM
Data Source(s)
World Wildlife Fund and ArcWorld™,
ESRI, U.S. Department of Interior,
National Park Service

Global 200—The World's Biologically Outstanding Ecoregions: Terrestrial, Freshwater, and Marine

The Global 200 map shows ecoregions that are outstanding examples of each major terrestrial, freshwater, and marine habitat type from biogeographic realms around the world. It targets biologically outstanding ecoregions that deserve greater conservation attention. With a focus on these more than 200 ecoregions, the broadest range of the world's species and habitats can be conserved while fostering the persistence of the ecological processes that maintain the web of life.

Ecoregions, rather than countries, are the units of conservation for this analysis because patterns of biodiversity do not reflect political boundaries. This is the first comprehensive effort to elevate outstanding examples of freshwater and marine ecosystems, as well as more familiar terrestrial ecosystems. This map synthesizes the results of intensive regional analyses of biodiversity across five continents. It was completed in collaboration with hundreds of regional experts and with hundreds of extensive literature reviews conducted by the World Wildlife Fund. The selection criteria were species richness, levels of endemism, taxonomic uniqueness, unusual ecological or evolutionary phenomena, and global rarity of the major habitat type. The Global 200 map was created in ArcInfo 8 using the text halo feature of ArcMap to label the ecoregion numbers. The transparency feature was used to enhance the marine ecoregions over bathymetry.

Joshua Tree National Park

ArcInfo 8 with ArcMap was used to create this map showing the various characteristics of Joshua Tree National Park. ArcMap helped to enhance the aesthetic quality of the map by generating a raster transparency of a digital elevation model draped over hill shade to display the majestic terrain of the park.

Possessing a rich human history and a pristine natural environment, Joshua Tree National Park spans the transition between the Mojave and Colorado deserts of Southern California. The characteristics of the two large ecosystems of these deserts are determined primarily by elevation. Below 3,000 feet, the Colorado Desert encompasses the eastern part of the park and features natural gardens of creosote bush, ocotillo, and cholla cactus.

The higher, moister, and slightly cooler Mojave Desert provides the special habitat the Joshua tree requires. Joshua tree forests occur in the western half of the park, which also includes some of the most interesting geologic displays found in California's deserts.

Joshua Tree National Park

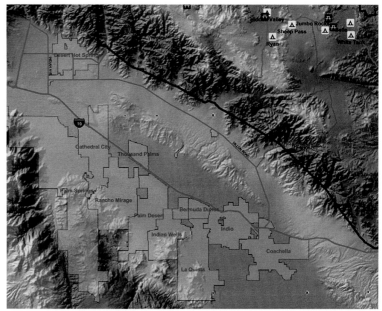

Juvenile Criminal Activities—The Spatial Extent

Police Beat Offense Frequency

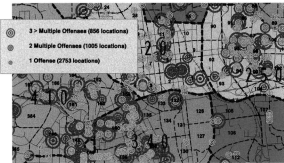

Fire Demand Zone (FDZ) Offense Frequency

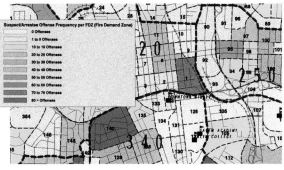

Juvenile Criminal Extent T1500-2100 hours

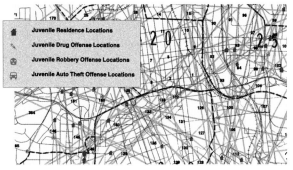

Juvenile Offense and Residence Locations

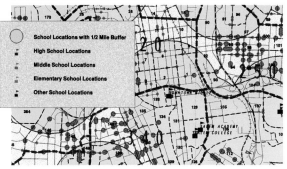

Strategic Approaches to Community Safety in Winston-Salem, North Carolina

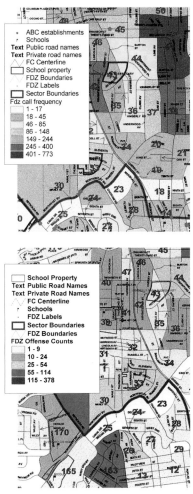

Police Beat Offense Frequency (1995–1998)

The victimized as well as the juvenile suspect and arrestee offense frequencies by police beat are shown here. Specific police beats are shown here to have a high frequency of instances of juvenile crime.

Fire Demand Zone (FDZ) Offense Frequency (1995–1998)

The FDZ data set derives both the fire department's fire-home territory and the police department's police beat data sets. High juvenile crime offense FDZ areas are located in specific areas of the city.

Juvenile Criminal Extent T1500–2100 hours (1995–1998)

Residences of the juveniles are linked by the type of offense, which is indicated by crime symbology, and are connected to the offense location by different line color symbology according to the type of offense.

Juvenile Offense and Residence Locations (1995–1998)

Offense locations and juvenile residences related to the offenses are shown. School location is overlaid in a .5-mile buffer and indicates that approximately 53 percent of the juvenile residences and approximately 58 percent of the juvenile offenses occur within this buffer area.

Strategic Approaches to Community Safety in Winston–Salem, NC

The thematic map in the upper left shows the frequency of juvenile offense involvement by FDZ for 1998. The location, role of juvenile, involvement frequency, race, sex, time of day, and day of week were used for further analysis. The map on the bottom left shows calls for service for specific crimes, liquor establishments, and schools. The spatial relationship function in ArcView GIS 3.1 was used to find all schools within 500 feet of liquor establishments.

City of Winston-Salem
Winston-Salem, North Carolina

By Tim Lesser

Contact
Tim Lesser
timl@ci.winston-salem.nc.us

Software
ArcInfo 7.2.1
Hardware
Sun Ultra 1 workstation
Printer
HP DesignJet 750C+
Data Source(s)
City of Winston-Salem Information Systems, City of Winston-Salem Police Department, and Forsyth County

(bottom left)
City of Winston-Salem
Winston-Salem, North Carolina

By Julia Conley

Contact
Julia Conley
juliac@wspd.org

Software
ArcView GIS 3.1, Visual FoxPro Version 6.0, and Crystal Reports Version 7.0
Hardware
Sun Ultra 4 workstation and Pentium PC
Printer
HP DesignJet 755CM
Data Source(s)
City of Winston-Salem/Forsyth County, Winston-Salem Police Department, Winston-Salem Fire Department, Winston-Salem Information Services Department, Forsyth County Assessor's Office, and North Carolina Alcohol Beverage Control Commission

San Diego Police Maps

San Diego Police Department
San Diego, California

*By Deena Bowman-Jamieson,
Matthew Turner, and Chad Yoder*

Contact
Deena Bowman-Jamieson
zgb@sdpd.sannet.gov

Software
ArcInfo 7.2.1, ArcPress, ArcView
GIS 3.1, Spatial and Temporal
Analysis of Crimes
Hardware
IBM RS6000 and Compaq Deskpro
Pentium III
Printer
HP DesignJet 3500CP
Data Source(s)
SanGIS, Automated Regional Justice
Information System (ARJIS), and San
Diego Police Department

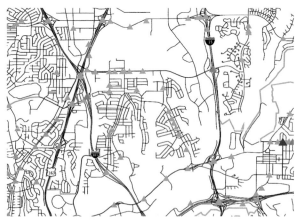

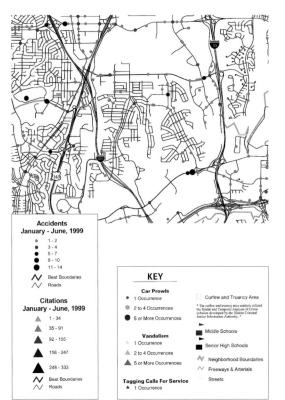

**Accidents
January - June, 1999**
- 1 - 2
- 3 - 4
- 5 - 7
- 8 - 10
- 11 - 14
N Beat Boundaries
N Roads

**Citations
January - June, 1999**
- 1 - 34
- 35 - 91
- 92 - 155
- 156 - 247
- 248 - 333
N Beat Boundaries
N Roads

KEY

Car Prowls
- 1 Occurrence
- 2 to 4 Occurrences
- 5 or More Occurrences

Vandalism
- 1 Occurrence
- 2 to 4 Occurrences
- 5 or More Occurrences

Tagging Calls For Service
- 1 Occurrence

☐ Curfew and Truancy Area
* The curfew and truancy area analysis utilized the Spatial and Temporal Analysis of Crime software developed by the Illinois Criminal Justice Information Authority.*

- ■ Middle Schools
- ■ Senior High Schools
- N Neighborhood Boundaries
- N Freeways & Arterials
- Streets

Excessive Truancy and Curfew Violation Areas and Selected Property Crimes, January–March 1999—This map, constructed using the Spatial and Temporal Analysis of Crimes (STAC) software, identifies target areas around the largest concentration of locations for curfew and truancy violations. Car prowls, vandalism, and tagging calls for service are often attributed to younger populations, and the map shows that many crimes overlap target areas, indicating that officers may be able to curtail two different types of problems by concentrating in one area of interest.

Traffic Accident and Speed Related Citation Locations, January–June 1999—This map exemplifies the type of information useful for evaluating the responsiveness of police personnel to traffic incidents. The map shows traffic accidents and citations for speed-related incidents in a selected area of San Diego. Officers seeing an abundance of collisions along with few citations in a particular part of town can shift their focus to the problem area.

Proximity of Firearm Incidents to Stolen Firearm Locations, January–June 1999—This map shows locations where firearms were stolen with concentric buffers a tenth of a mile around these locations. Violent crimes involving a firearm and calls for service for discharging a firearm were overlaid on the concentric buffers. The map provides officers with information about areas that may need attention, specifically in dealing with the reduction of firearm incidents.

Residential Burglaries Suspect to Victim Locations—On this map, straight-line distances were calculated between the suspect(s) and victim locations to give officers a better idea about the geographic nature of the crimes. This valuable information gives officers an idea of where to begin searching for suspects when new cases occur. Officers can overlay more information on the map such as known loitering areas or narcotics houses.

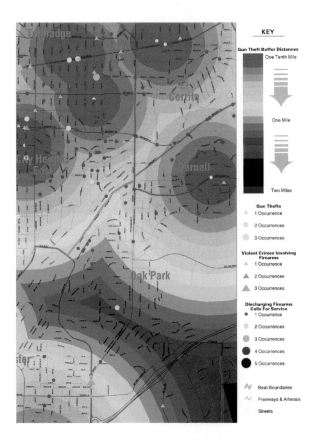

KEY

Gun Theft Buffer Distances
One Tenth Mile
One Mile
Two Miles

Gun Thefts
- 1 Occurrence
- 2 Occurrences
- 3 Occurrences

Violent Crimes Involving Firearms
- 1 Occurrence
- 2 Occurrences
- 3 Occurrences

Discharging Firearms Calls For Service
- 1 Occurrence
- 2 Occurrences
- 3 Occurrences
- 4 Occurrences
- 5 Occurrences

N Beat Boundaries
N Freeways & Arterials
Streets

U.S. Bureau of Reclamation MPGIS Projects in Water Resources and Emergency Response

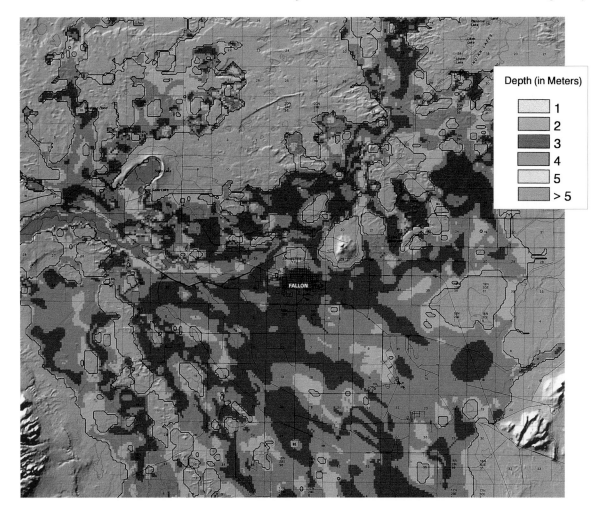

Depth (in Meters)

	1
	2
	3
	4
	5
	> 5

U.S. Bureau of Reclamation
Sacramento, California

*By Dave Hansen, Thomas Heinzer,
Ellie Robbins, Michael Sebhat, and
Barbara Simpson*

Contact
Michael Sebhat
msebhat@mp.usbr.gov

Software
ArcInfo 7.2.1, ArcView GIS 3.1, and
T2View
Hardware
HP UNIX workstation
Printer
HP DesignJet 2500CP
Data Source(s)
U.S. Geological Survey

This Lahontan Dam breach simulation was
performed using the one-dimensional NWS
DAMBRK and two-dimensional MIKE21 hydraulic
models. The MIKE21 simulation was visualized
using ArcInfo utilities.

Presented on this map is a simulation of the
maximum inundation over time resulting from the
failure of the Lahontan Dam with the reservoir
at full capacity. The inundated areas shown on
the map based on the results of this study reflect
extremely rare events. The results depicted on the
inundation map approximate the flood boundaries
and condition resulting from dam failure. These
boundaries could be more or less severe than
indicated.

This study has not yet been formally completed and
reviewed.

Tornadic Activity in Wichita, Kansas: Impact Assessment and Disaster Preparedness

City of Wichita, GIS
Wichita, Kansas

By Jan Keathley, Mike Kollmeyer, and Lori Wilkerson

Contact
Lori Wilkerson
Wilkerson_L@ci.wichita.ks.us

Software
ArcInfo 7.0.4 and ArcView GIS 3.1
Hardware
DEC Alpha workstation
Printer
HP DesignJet 2500CP
Data Source(s)
Wichita Water and Sewer Department, Federal Emergency Management Agency (FEMA), National Weather Service (NWS), Wichita Central Inspection, Sedgwick County GIS, U.S. Department of Commerce, National Oceanic and Atmospheric Administration, U.S. Bureau of the Census, ESRI sample data, http://mattdennis.com/skywarn, and City of Wichita GIS

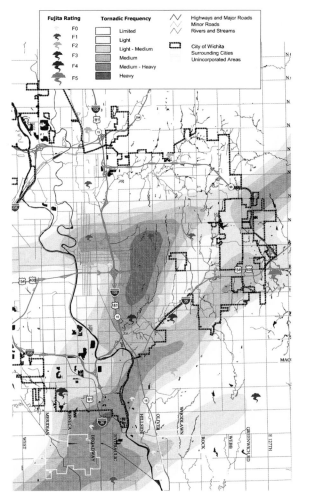

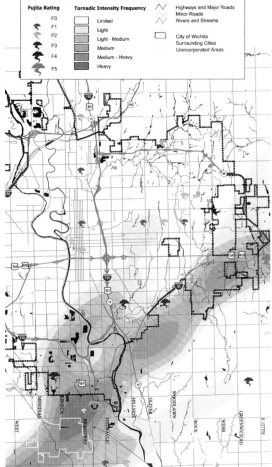

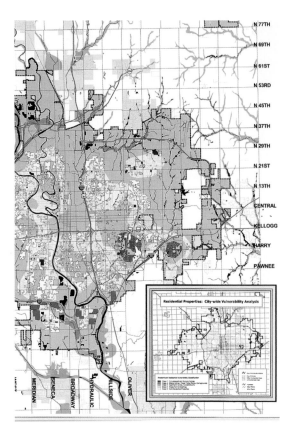

This project was developed by Wichita's GIS staff to support relief efforts for the tornado disaster of May 1999 and to assist the city in preparing for future tornadic activity and related severe weather conditions.

GIS is helping to identify the location and extent of damage to homes, to determine reconstruction policies, and to prioritize plans for rebuilding based on environmental factors and structural characteristics. GIS is also enabling the creation of a more disaster-resistant economy and community.

A tornadic vulnerability analysis was performed to identify residential structures that do not have basements and may not provide adequate protection to residents during severe windstorms. The need for on-site shelter was assigned a priority rating based on the structure type and characteristic of the residents. A composite analysis was performed to identify more generalized areas throughout the city that contain a predominance of highly vulnerable residential structures and a high population density.

Historical tornadoes were mapped as paths extending between known start and end locations. Tornado paths less than one-half mile in length were mapped as a single point. Using a grid cell size of 300 square feet, the storm paths were generalized and ranked to reflect relative tornado intensity based on the Fujita scale rating system.

Seismic Vulnerability of the Puget Sound Region

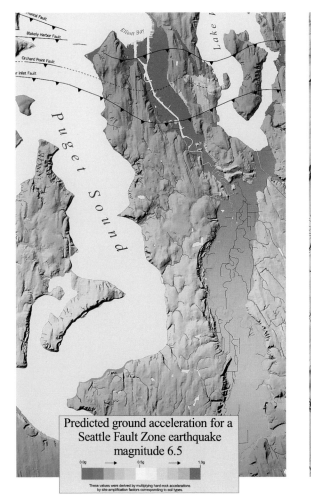

Predicted ground acceleration for a Seattle Fault Zone earthquake magnitude 6.5

King County GIS
Seattle, Washington

By Michael Jenkins

Contact
Michael Jenkins
michael.jenkins@metrokc.gov

Software
ArcInfo 7.1.2, ARC GRID, ArcView GIS 3.1, and ArcView Spatial Analyst
Hardware
Pentium-based PC
Printer
HP DesignJet 2500CP
Data Source(s)
U.S. Geological Survey, King County GIS, and Pierce County GIS

King County is the largest county in Washington, home to half of the state's residents and predominately urban, with more than 75 percent of its residents living in the incorporated areas near Seattle. Other communities in the county are sufficiently isolated by Lake Washington and have a separate identity.

King County is highly susceptible to earthquakes, floods, landslides, wildland fires, and winter storms. The county is at risk for a Great Cascadia Subduction Zone earthquake, which could cause extensive damage. Currently operating under Uniform Building Codes (UBC) Seismic Zone 4, the county is extremely disaster prone with eight declared disasters within the last 10 years. The declarations were principally the result of flooding and winter storms and resulted in directing more than $65 million to the county. These funds have been used to reduce the vulnerability of the county's infrastructure and homes to flood damage.

King County is participating in the National Flood Insurance Program's Community Rating System (CRS). It is one of only two Class 6 CRS communities in the country. Early on, the county established strong flood mitigation measures including an ordinance that substantially exceeded the minimum federal regulations. In addition, the county established a drainage utility that provided tax monies for mitigation and prepared flood studies using Federal Emergency Management Agency guidelines and support.

FEMA's Project Impact—Building Disaster Resistant Communities helps communities protect themselves from the devastating effects of natural disasters by taking actions that dramatically reduce disruption and loss.

King County surrounds Seattle, which was one of FEMA's pilot Project Impact communities. The public and private partnerships forged through the Seattle Project Impact initiative will assist King County in the development of their Project Impact initiative. It is anticipated that the county's and the city's Project Impact initiatives will complement each other.

This map was made to support the Project Impact initiative in King and Pierce counties. The purpose was to illustrate the vulnerability of the transportation system in the region. Geologists at the U.S. Geological Survey supplied values representing predicted ground acceleration and susceptibility to soil liquefaction. They were combined with hillside data and processed in ARC GRID and ArcView Spatial Analyst for the final map layouts.

Spatial Interpretation Laboratory, California Institute of Technology

A TriNet Shake Map

California Institute of Technology
Pasadena, California

*By Danny H. Natawidjaja, Kerry Sieh,
Tony Soeller, and David J. Wald*

Contact
Kerry Sieh
sieh@gps.caltech.edu

Software
ArcInfo 7.2.1 and ArcView GIS
Hardware
Sun workstation, PC, and Mac

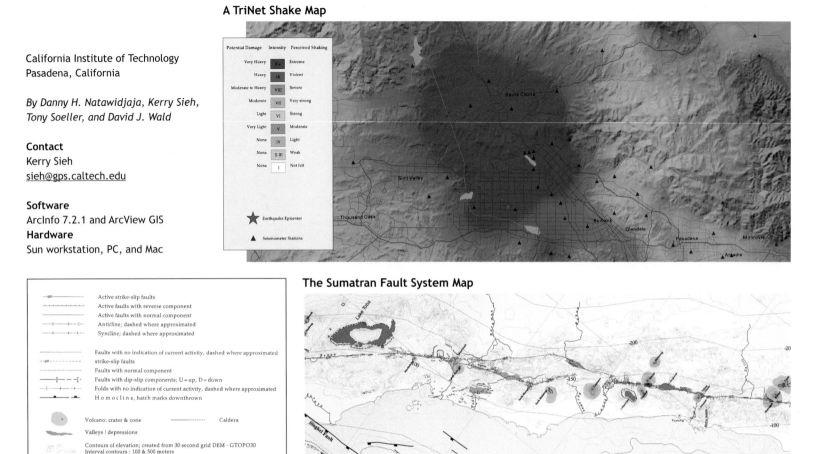

The Sumatran Fault System Map

A TriNet Shake Map

The research and development efforts of the TriNet project (California Institute of Technology [Caltech], California Division of Mines and Geology [CDMG], and U.S. Geological Survey [USGS]) have created ShakeMaps for earthquakes in Southern California. Triggered by any significant earthquake in Southern California, the maps are generated automatically and are made available on the Internet (www.trinet.org/shake.html) within several minutes of an earthquake.

Data from the USGS/Caltech stations is acquired in real time using a variety of digital telemetry methods. CDMG stations are near real time, using an automated dial-up procedure developed for use with telephone communications. Initial maps are made (within five minutes) with the real-time component of TriNet, but they are updated automatically as more data is acquired from CDMG.

A detailed description of the shaking over the Southern California region requires interpolation of the nonuniformly spaced, observed ground motions. Peak ground motions are estimated in areas of sparse station coverage by fitting an attenuation relationship to the data and scanning the parameter space for the best latitude, longitude, and magnitude of the strong motion "centroid."

The Sumatran Fault System Map

Caltech created the Sumatran Fault System map to examine in detail the geometry of the Sumatran Fault in Indonesia. Such analysis has increased the understanding of the relationship of the fault's structure and position in relation to events such as volcanism and earthquakes. The full version of the map is available as PostScript and ArcInfo files at http://www.scecdc.scec.org/geologic/sumatra.

The Sumatran Fault is a 1,900-kilometer-long trench parallel strike-slip structure between the Eurasian and Indian/Australian plates. The detailed map of the fault, compiled from topographic maps and stereographic aerial photographs, shows that unlike many other great strike-slip faults the Sumatran Fault is highly segmented. The influence of these stepovers on historical seismic source dimensions suggests that the dimensions of future events will also be constrained by fault geometry.

Colorado Springs Utilities Enterprisewide GIS

Colorado Springs Utilities
Colorado Springs, Colorado

*By Betty Jo Alegria, Steve Clark,
Larry Gillette, David Krenick,
Randy Scott, Paul Stella, and
Margaret Thompson*

Contact
Jim Van Riper
jvanriper@csu.org

Software
ArcView GIS and ARCPLOT
Hardware
Sun SPARC workstation and Dell
Pentium laptop
Data Source(s)
Internal data

Several map products from Colorado Springs Utilities depict
the enterprisewide utilization of GIS there. These plots
ranged from standard map book plots used by engineering,
construction, and office personnel to custom plots created
for specific projects. The maps all share a common set of
land base data created and maintained by a central group
and utility-specific data created and maintained by those
respective departments.

Greater Cairo Utility Information

Greater Cairo Utility Data Center
Cairo, Egypt

By Dalal Abdul-Mageed

Contact
Dalal Abdul-Mageed
gcudc@idsc.gov.eg

Software
ArcInfo 7.2.1
Hardware
Sun Ultra workstation
Data Source(s)
Field surveying

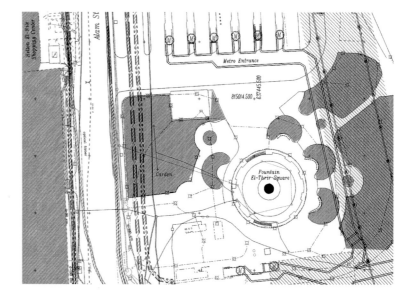

Maps of two squares in Cairo depict the enormous extent of underground utilities. Maps such as these are important reference tools that help guide construction digging to avoid underground hazards.

The map information includes street names and all pertinent utility information such as type, size, and depth. Each utility is depicted with a different color.

Ecological Land Types of the Hudson Valley Section

Kittatinny -Schawangunk Mountains & Ridges Subsection

Kittatinny Limestone Valley Subsection

ELT Codes

Moisture Fertility

- 11
- 12
- 13
- 16
- 21
- 22
- 23
- 32
- 33
- 41
- 42
- 43
- 51
- 52
- 53
- 54
- 63
- 64
- 80
- 86
- 99

Ecological land types (ELTs) are geographic units with physical characteristics including landscape position, soil fertility, water availability, elevation, slope, aspect, and relative illumination. The New Jersey Forest Service used ARC GRID to model the ELT level of the ECOMAP classification system, which was developed by the U.S. Forest Service and others. The map shows ELTs of one section of New Jersey.

Each ELT (referenced with a two-digit code) identifies underlying characteristics of the site and consequently reflects the potential plant communities it may support. Text descriptions for each ELT describe the findings of field data and suggest potential successive patterns of natural vegetation. This project is facilitating long-term forest management planning and restoration efforts across the landscape. The project is scheduled for publication.

New Jersey Forest Service
  Department of Environmental
  Protection
Trenton, New Jersey

By Craig Coutros, James Dunn, and Dave MacFarlane

Contact
Craig Coutros
craigc@gis.dep.state.nj.us

Software
ArcInfo 7.2 and ARC GRID
Hardware
Sun Ultra workstation
Printer
HP DesignJet 1055CM Inkjet
Data Source(s)
Vegetation information collected and analyzed on a statewide basis by forest ecologists over a five-year period and U.S. Geological Survey digital elevation model

Coastal Mapping in Support of the Southwest Washington Coastal Erosion Study

Southwest Washington Coastal
 Erosion Study
Olympia, Washington

*By R.C. Daniels, S. Eykelhoff,
R.H. Huxford, D. McCandless, and
B. Voigt*

Contact
Richard Daniels
rdan461@ecy.wa.gov

Software
ArcInfo 7.1, ArcView GIS 3.1, and
ERDAS IMAGINE Version 8.3
Hardware
Sun Ultra 1 workstation, Sun SPARC 20
workstation, and Windows 95 PCs
Data Source(s)
National Aeronautics and Space
Administration/National Oceanic and
Atmospheric Administration ALICE
Project and U.S. Geological Survey
Southwest Washington Coastal
Erosion Study

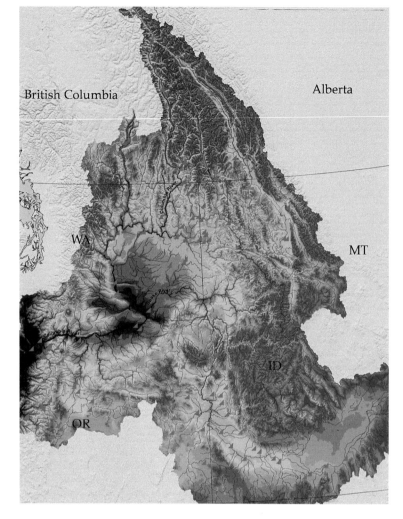

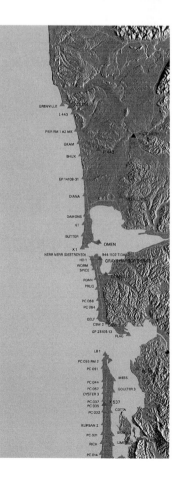

The Southwest Washington Coastal Erosion Study is a joint federal, state, and local partnership led by the U.S. Geological Survey Marine and Coastal Geology Program and the Washington State Department of Ecology. The study seeks to document historical and current shoreline position and change rates within the Columbia River littoral cell (CRLC) and to assess and predict the effects that variations in sediment supply to the system might have on the position of the shoreline.

The CRLC extends from Tillamook Head, Oregon, to Point Grenville, Washington. This 165-kilometer-long region historically received most of its sand from the Columbia River, which is about 2,000 kilometers long and drains more than 675,000 square kilometers of Washington, Oregon, Idaho, and British Columbia. Twelve major dams have been constructed along the river since 1990 and have contributed to substantially decreased amounts of sand reaching the coast. This decrease in sediment supply, coupled with dredging efforts (from a mean channel depth of 25 feet in 1914 to more than 45 feet in 1990) at the mouth of the Columbia River, could be responsible for the increasing erosion rates within the CRLC.

As part of this study, the Department of Ecology has developed a coastal GIS for the region. This GIS is unique in that it is based on a high-accuracy reference network developed in conjunction with the National Geodetic Survey. All the data layers within the GIS, including orthophoto mosaics, vector shorelines, and ground global positioning system survey work, are tied into this network. The ecology department has used this GIS to develop hundreds of map products to support researchers and coastal managers within the region. Examples of these products include shoreline position mapping and topographic mapping for wetland delineation.

Water Quality Analysis in the Eastern Iowa Basins Study Unit

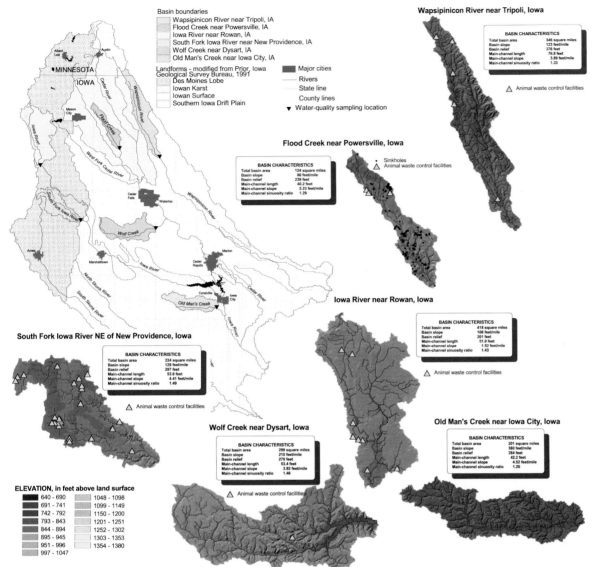

Basin boundaries
- Wapsipinicon River near Tripoli, IA
- Flood Creek near Powersville, IA
- Iowa River near Rowan, IA
- South Fork Iowa River near New Providence, IA
- Wolf Creek near Dysart, IA
- Old Man's Creek near Iowa City, IA

Landforms - modified from Prior, Iowa Geological Survey Bureau, 1991
- Des Moines Lobe
- Iowan Karst
- Iowan Surface
- Southern Iowa Drift Plain

- Major cities
- Rivers
- State line
- County lines
- ▼ Water-quality sampling location

Wapsipinicon River near Tripoli, Iowa

BASIN CHARACTERISTICS
Total basin area	346 square miles
Basin slope	123 feet/mile
Basin relief	376 feet
Main-channel length	76.8 feet
Main-channel slope	3.89 feet/mile
Main-channel sinuosity ratio	1.33

△ Animal waste control facilities

Flood Creek near Powersville, Iowa

BASIN CHARACTERISTICS
Total basin area	124 square miles
Basin slope	86 feet/mile
Basin relief	239 feet
Main-channel length	40.2 feet
Main-channel slope	5.23 feet/mile
Main-channel sinuosity ratio	1.29

· Sinkholes
△ Animal waste control facilities

Iowa River near Rowan, Iowa

BASIN CHARACTERISTICS
Total basin area	418 square miles
Basin slope	106 feet/mile
Basin relief	201 feet
Main-channel length	51.9 feet
Main-channel slope	1.52 feet/mile
Main-channel sinuosity ratio	1.43

△ Animal waste control facilities

South Fork Iowa River NE of New Providence, Iowa

BASIN CHARACTERISTICS
Total basin area	224 square miles
Basin slope	129 feet/mile
Basin relief	297 feet
Main-channel length	53.8 feet
Main-channel slope	4.41 feet/mile
Main-channel sinuosity ratio	1.49

△ Animal waste control facilities

Wolf Creek near Dysart, Iowa

BASIN CHARACTERISTICS
Total basin area	299 square miles
Basin slope	210 feet/mile
Basin relief	270 feet
Main-channel length	53.4 feet
Main-channel slope	3.92 feet/mile
Main-channel sinuosity ratio	1.46

△ Animal waste control facilities

Old Man's Creek near Iowa City, Iowa

BASIN CHARACTERISTICS
Total basin area	201 square miles
Basin slope	380 feet/mile
Basin relief	284 feet
Main-channel length	42.2 feet
Main-channel slope	4.52 feet/mile
Main-channel sinuosity ratio	1.20

ELEVATION, in feet above land surface
640 - 690	1048 - 1098
691 - 741	1099 - 1149
742 - 792	1150 - 1200
793 - 843	1201 - 1251
844 - 894	1252 - 1302
895 - 945	1303 - 1353
951 - 996	1354 - 1380
997 - 1047	

U.S. Geological Survey
Iowa City, Iowa

By Kymm Akers

Contact
Kymm Akers
kkakers@usgs.gov

Software
ArcInfo 7.2.1 and ArcView GIS 3.1
Hardware
Dell Dimension XPS R350
Printer
HP DesignJet 2500CP
Data Source(s)
U.S. Geological Survey and Iowa Department of Natural Resources, Geological Survey Bureau

Data collection activities in the Eastern Iowa Basins began in September 1995 as part of the National Water Quality Assessment Program. The long-term goals of this program are to describe the status of and trends in the quality of a large, representative part of the nation's surface and groundwater resources and to identify the major factors that affect the quality of the resources. The Eastern Iowa Basins study unit, covering approximately 19,500 square miles, was selected as an important hydrologic system representative of an agricultural area in the Midwest. The study unit comprises the watershed of four major rivers in Iowa that flow in a southeasterly direction toward eventual discharge into the Mississippi River.

In addition to the spatial relations between six watersheds and physiographic condition, this map shows the location of animal waste control facilities that could potentially impact water quality. A GIS was used in generating a hydrologically enforced 1:100,000-scale digital elevation model that was then used in conjunction with stream data to further quantify basin characteristics unique to each watershed. These basin characteristics, along with many other ancillary data sets, are used to identify factors affecting water quality in the Eastern Iowa Basins study unit.

Billings, Montana, 1996 Roadway Motor Vehicle Carbon Monoxide Emissions Modeling Project

Montana State Library
Helena, Montana

*By Cyra Cain, Jim Carlin, Kristina
Gurrieri, Dave Highness, and
David J. Schnittgen*

Contact
David Highness
dhighness@state.mt.us

Software
ArcInfo 7.2.1, Solaris Version 2.5.1,
and Patch Version 1
Hardware
Sun Ultra 1 workstation
Printer
HP DesignJet 650C
Data Source(s)
ESRI U.S. street database, Montana
Department of Transportation roads
data, and Montana Department of
Environmental Quality carbon
monoxide data

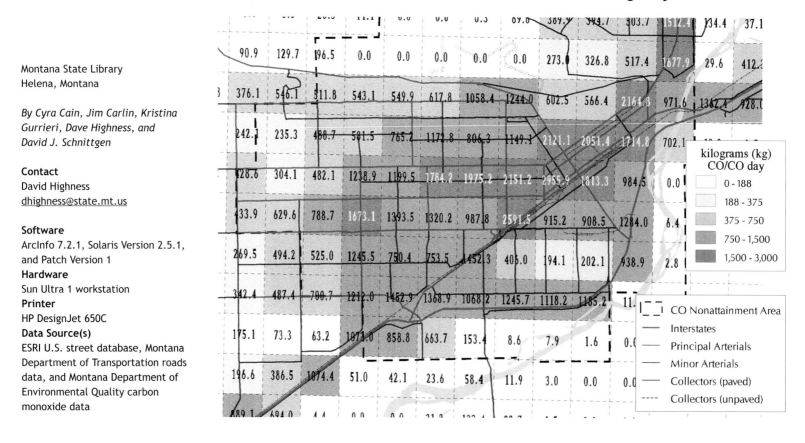

This project is part of an ongoing effort by the Department of Environmental Quality to reduce carbon monoxide emission in Montana's cities and to reclassify those cities from nonattainment status to attainment status. The poster represents the vehicular carbon monoxide modeling project for Billings, Montana. It presents the objectives, methods, and results of the production of a grid of daily carbon monoxide emissions created by automobiles in Billings, Montana, and surrounding areas.

The project objective was to develop a GIS database to study motorized vehicle emissions by grid cell and traffic flow in the greater Billings area. The project represents the emissions of motorized vehicles as part of the greater Billings area carbon monoxide emissions inventory.

Average daily traffic values from the Montana Department of Transportation were linked to a commercial roads coverage. The roads coverage was then split by a one-kilometer grid using the ArcInfo identity command. This GIS operation enabled the roads to be summarized by grid cell and vehicle kilometers traveled per day per grid cell. Roads were further summarized by road classification. Using these summarized values the Department of Environmental Quality was able to model average daily carbon monoxide emissions per grid cell.

This project is the result of an interagency cooperative effort between the Department of Environmental Quality, the Montana Department of Transportation, and the Natural Resource Information System of the Montana State Library. The state agencies involved in this effort are especially proud of this map as it is a result of multiagency cooperation.

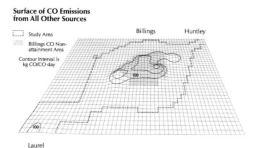

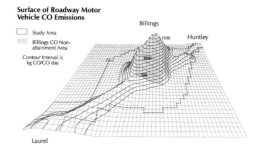

Surface of Roadway Motor Vehicle CO Emissions

Surface of CO Emissions from All Other Sources

Potential Acoustic Impacts of Rocket Engine Tests

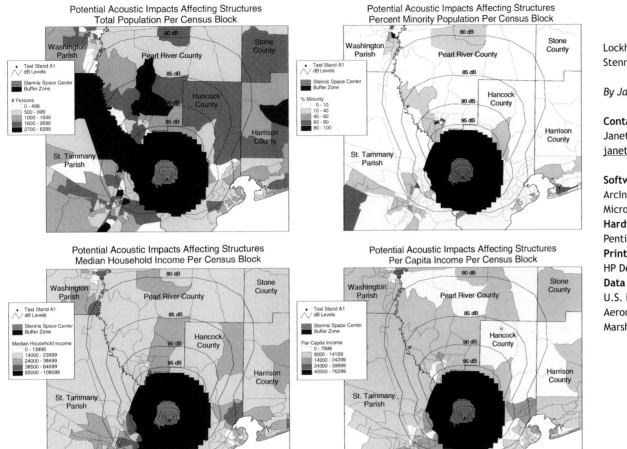

Potential Acoustic Impacts Affecting Structures
Total Population Per Census Block

Potential Acoustic Impacts Affecting Structures
Percent Minority Population Per Census Block

Potential Acoustic Impacts Affecting Structures
Median Household Income Per Census Block

Potential Acoustic Impacts Affecting Structures
Per Capita Income Per Census Block

Lockheed Martin Space Operations
Stennis Space Center, Mississippi

By Janette Lovely

Contact
Janette Lovely
janette.lovely@ssc.nasa.gov

Software
ArcInfo 7.2.3, ArcView GIS 3.0a, and
Microsoft PowerPoint 97
Hardware
Pentium Windows NT
Printer
HP DesignJet 755CM
Data Source(s)
U.S. Bureau of Census and National
Aeronautics and Space Administration
Marshall Space Flight Center

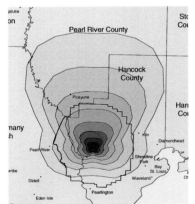

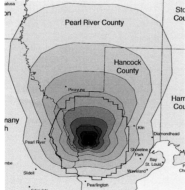

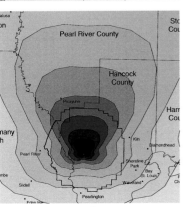

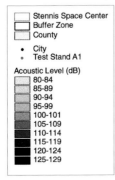

Stennis Space Center
Buffer Zone
County

• City
• Test Stand A1

Acoustic Level (dB)
80-84
85-89
90-94
95-99
100-101
105-109
110-114
115-119
120-124
125-129

The John C. Stennis Space Center (SSC) Environmental GIS is a database for the SSC and the surrounding eight-county area. The GIS contains acoustic models from rocket engine tests as well as remote sensing and ancillary data. The data was gathered from many sources including local, county, state, and U.S. government and private agencies. The main purpose for the GIS is to aid SSC public outreach and site management.

The maps detail the acoustic model for the RD-180 engine along with demographic information contained in the GIS. These maps show how the GIS has assisted SSC in noise pollution modeling, especially as it relates to environmental justice concerns.

The next set of maps illustrates the potential acoustic impacts of three different types of rocket engines on the surrounding area, showing the difference in acoustic levels as distance from the test stand increases. The models measure the low frequency noise affecting structures. Acceptable levels are 120 decibels or lower outside of the National Aeronautics and Space Administration buffer zone.

Multitemporal Landsat Thematic Mapper Satellite Imagery for Agricultural Land Use Classification Within the Mermentau River Basin

Louisiana Department of
 Environmental Quality (LDEQ) GIS
 Center
Baton Rouge, Louisiana

By Brad Mooney

Contact
Paul Zundel
paul_z@deq.state.la.us

Software
ArcInfo 7.1.2, ArcView GIS 3.1.1,
ERDAS IMAGINE Version 8.3.1, and
Adobe Photoshop Version 5.0
Hardware
Windows NT TD workstations
Printer
DisplayMaker 6000
Data Source(s)
Landsat Thematic Mapper satellite
imagery (acquired on April 20, 1998;
August 26, 1998; October 13, 1998;
and December 16, 1998), 1995
National Aeronautics and Space
Administration high-altitude infrared
aerial photography, 1999 LDEQ low-
altitude aerial photography and
ground photography, 1999 LDEQ
Basin-Subsegment data, and 1998
Bayou Plaquemine Brule (LDEQ Basin-
Subsegment #050201) ground truth
data

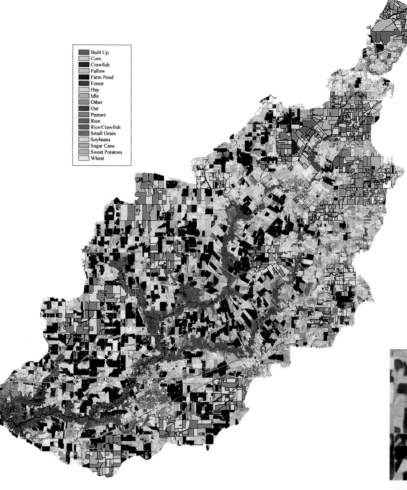

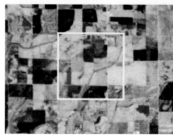

GIS has played a critical role in the efforts of the Louisiana Department of Environmental Quality (LDEQ) to map and analyze watersheds within Louisiana. The department uses multiple data sources including satellite imagery. This layout shows the utility of multitemporal Landsat Thematic Mapper satellite imagery for classifying agricultural land use within the Mermentau River basin of southwestern Louisiana.

The study area corresponds to the boundary of LDEQ Basin-Subsegment #050201, also known as Bayou Plaquemine Brule. Bayou Plaquemine Brule is a subbasin watershed located within the Mermentau River basin. The layout was created to display the complexity of agricultural land use practices within southwestern Louisiana, while detailing a methodology for mapping such land use.

This project has been a multiagency endeavor. The LDEQ Nonpoint Source Unit and GIS Center worked closely with the Louisiana Department of Agriculture and Forestry (Office of Soil and Water Conservation), the U.S. Department of Agriculture (Natural Resources Conservation Service and Farm Service Agency), and the Acadia and St. Landry Parish Soil and Water Conservation districts throughout the course of the project.

Example Using ArcView 3D Analyst and ArcView Image Analysis to Display LDEQ Air Modeling Data from Two Emission Sites

Street Map

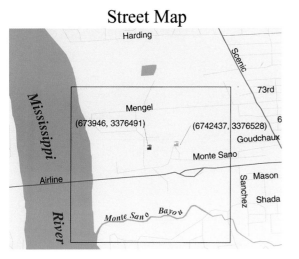

USGS 1:24000 DRG

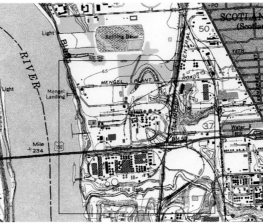

2D TIN Model of Emissions

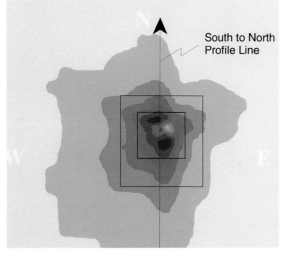

3D TIN Model of Emissions

Contoured Emissions

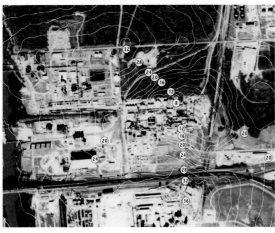

Land Use \ Contoured Emissions

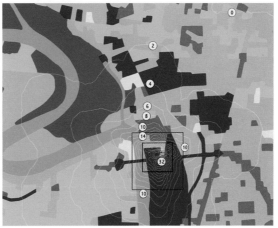

Louisiana Department of
  Environmental Quality (LDEQ) GIS
  Center
Baton Rouge, Louisiana

By Dan DuVal

Contact
Paul Zundel
paul_z@deq.state.la.us

Software
ArcView GIS 3.1.1, ArcView 3D
Analyst, and ArcView Image Analysis
Hardware
Windows NT TD workstations
Printer
DisplayMaker 6000
Data Source(s)
LDEQ modeled air quality data;
U.S. Geological Survey (USGS)
1:100,000-scale digital line graph;
USGS digital raster graphic; 1986
USGS land use generalized by U.S.
Environmental Protection Agency;
and January 1995 National
Aeronautics and Space Administration
infrared aerial photography

This LDEQ map is a montage of six displays depicting some of the pertinent information related to two air emission sites in Baton Rouge, Louisiana, such as major geographic features; contoured, profiled, and three-dimensional surface modeled emissions; and land use. This display was created to help LDEQ engineers better assess the impact of air emission sites on the surrounding community.

Calculation of the Spread (Dispersal) of Traffic Pollutants Based on Real Wind Field Modeling in Street Ravines of Cities and Town Centers

Engineering Office for Environmental
  Protection and Fluid Technology
Wettin, Germany

By Dr. Ing. Habil. Rainer Schenk

Contact
Rainer Schenk
ibsgmbh@compuserve.com

Software
ArcInfo 7.1.2
Hardware
IBM RS6000 workstation and PC
Pentium workstation
Printer
HP DesignJet
Data Source(s)
Municipal Administration of Neuss

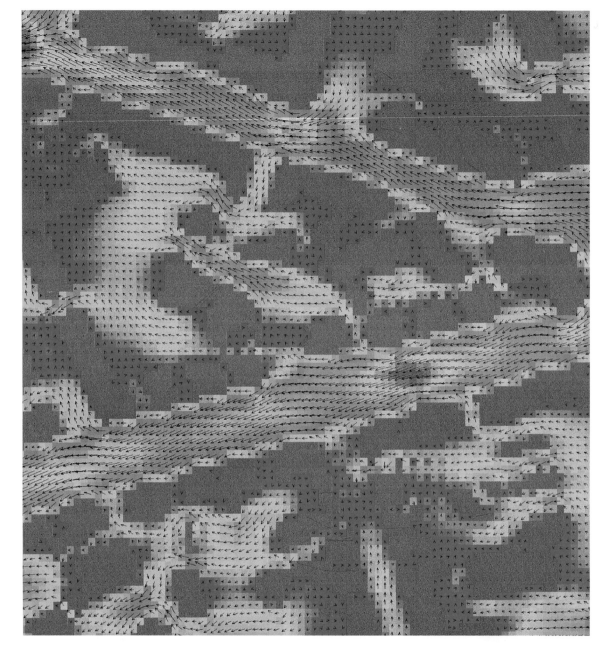

The calculation of the spread of air pollutants, especially in highly contaminated areas of heavily crowded (trafficked) cities and towns, is a very important task of environmental protection. The pollutant simulation is based on a realistic calculation of the wind fields in street ravines. The quality-of-spread calculations are determined largely by the degree of spatial resolution and the development structure of the area investigated.

The graphics show the results of three-dimensional wind field modeling. Here the speed vectors of some raster groups are represented including their dimensions and direction. The spread of air pollutants is characterized by discrete color gradations. The calculation and graphic presentation of the spread of traffic pollutants take into account real three-dimensional wind field modeling in cities and town centers.

The spread of traffic pollutants depends on the development structure as well as on meteorological conditions in the area investigated. Traffic emissions are represented as linear sources. Three-dimensional wind field modeling in cities and town centers was a precondition for calculating the spread of air pollutants. The speed vectors are represented including their dimensions and direction.

Ecomorphological Quality of Flowing Water in Hessen, Germany

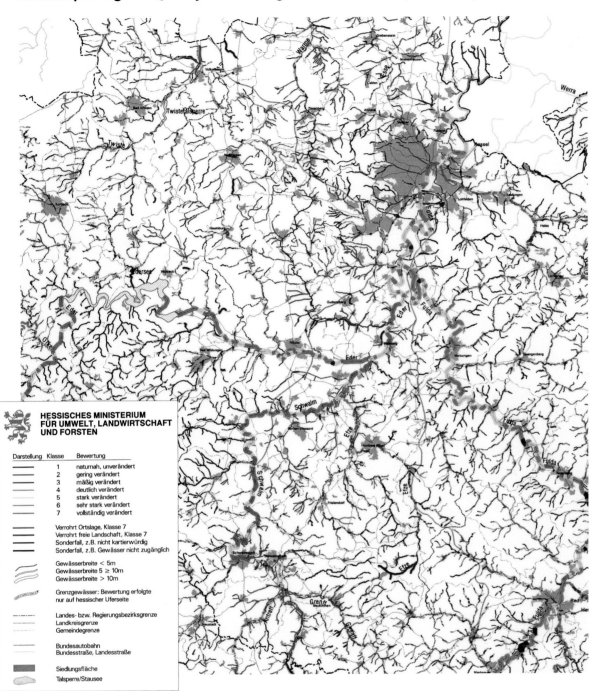

Ministry of Environment, Agriculture,
and Forestry
Wiesbaden, Hessen, Germany

*By Martin Blank, Rolf Feltens,
Dr. Stephan V. Keitz, and Frank Loy*

Contact
Dr. Stephan V. Keitz
s.keitz@mue.hessen.de

Software
ArcInfo 7.2.1
Hardware
Sun Enterprise
Printer
HP DesignJet 1055CM
Data Source(s)
ATKIS

47 | environmental management

The Hessen environmental ministry determined the ecomorphological quality of all streams and rivers in the County of Hessen, Germany, in order to support the sustainable development of flowing waters. The important morphological characteristics of more than 22,500 kilometers of rivers and streams were recorded and analyzed at each riverside between 1994 and 1999.

The results are documented in this 1:200,000-scale map. Additionally, more than 750 detailed map sheets at 1:10,000 scale depict the causes for the numerous ecological deficits with pictograms for every 100-meter section of water.

This information is accessible online at www.herasum.de.

Geology of Placitas, New Mexico

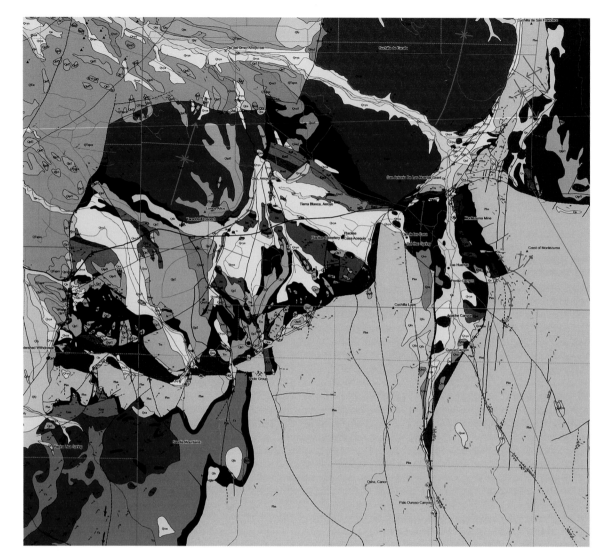

ESRI
Redlands, California

*By Charlie Frye, Prashant Hedao,
Mike Price, and Jaynya Richards*

Contact
Mike Price
mprice@esri.com

Software
ArcInfo 8, ARC GRID
Hardware
IBM PC
Printer
HP DesignJet 1055CM
Data Source(s)
New Mexico Bureau of Mines and
Mineral Resources

The Placitas map demonstrates the ability of ArcInfo 8 to generate, display, and plot complex surface geologic, topographic, and cultural information.

The Placitas, New Mexico, 7.5-minute quadrangle was mapped by field geologic staff of the New Mexico Bureau of Mines and Mineral Resources, Socorro, New Mexico. The Placitas quadrangle is located at the north end of the Sandia Mountains, approximately 25 miles northeast of Albuquerque, New Mexico.

Bureau field geologists collected polygonal outcrop geology, structural lines, and numerous geologic point data. Field geologic data was compiled at a scale of 1:12,000, then drafted and digitally compiled at 1:24,000 scale as ArcView GIS shapefiles by the bureau's cartographic staff. Digital data was provided to ESRI to aid ArcInfo 8 development.

The shapefiles and other data were imported into ArcInfo 8 to test thematic modeling, map production, and other features. Geologic polygons were thematically displayed using standard New Mexico Bureau of Mines stratigraphic and lithologic colors. Line themes, including faults and fold axes, were displayed with U.S. Geological Survey (USGS) standard symbology for 1:24,000-scale mapping. Geologic points, including strike, dip, foliation, and others, were also symbolized to USGS standards.

A 10-meter digital elevation model of the Placitas quadrangle was acquired from the USGS Geo Data Internet site and converted to an ARC GRID image. The ARC GRID and hill shade image are displayed above polygonal geology with a transparency that represents topographic relief. The vector data legend thematically shows feature type.

Map label points were downloaded from the USGS Geographic Names Information System site and used to label topographic and cultural features. Map names are generated and displayed using ArcInfo 8 labeling functionality. The U.S. Bureau of Mines Minerals Availability System data points show the location, type, and status of mineral deposits in the Placitas area.

Dominant Soil Orders and Suborders 1998—United States of America

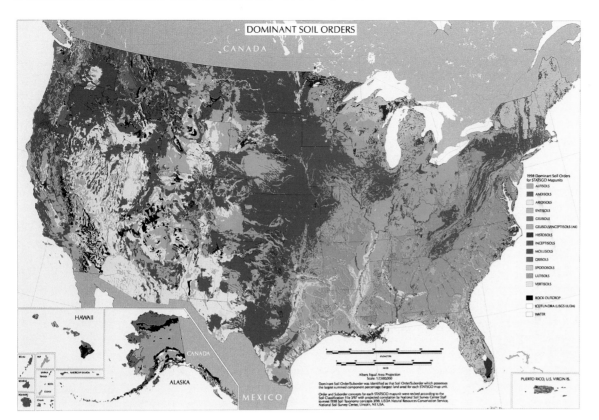

U.S. Department of Agriculture–
  Natural Resources Conservation
  Service
Lincoln, Nebraska

By Soil Survey Staff

Contact
Sharon Waltman
sharon.waltman@nssc.nrcs.usda.gov

Software
ArcInfo 7.2.1
Hardware
Sun Ultra 1 workstation
Printer
HP DesignJet 755CM
Data Source(s)
U.S. Department of Agriculture–
Natural Resources Conservation
Service, State Soil Geographic
Database National Collection, and
Correlated Soil Classification File

This map shows the geographic distribution of dominant soil orders and suborders according to the *1999 Soil Taxonomy* for the United States and those nations and territories serviced by the U.S. Department of Agriculture–Natural Resources Conservation Service. Soil order colors are reflected in the titles of 12 order inset maps. Each order inset map shows the distribution of respective suborders. This map was created for publication with the *1999 Soil Taxonomy Agricultural Handbook 436.*

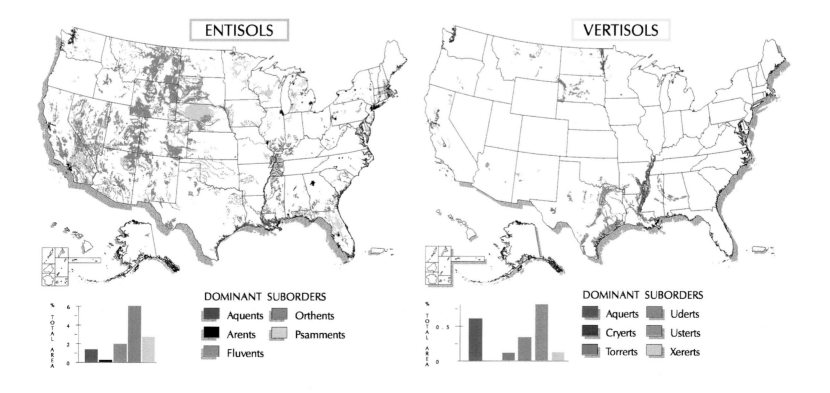

Seismicity Maps of the San Francisco and San Jose Quadrangles, 1967-1993

U.S. Geological Survey
Menlo Park, California

*By Ronan Mandel,
David Oppenheimer, and
Stephen Walter*

Contact
Steve Walter
swalter@usgs.gov

Software
ArcInfo 7.1.4, Adobe Illustrator
Version 5.0, and in-house mechanism
and plotting software
Hardware
Sun SPARC 20 workstation
Printer
Offset-printed
Data Source(s)
U.S. Geological Survey, Earthquake
Catalog of the Northern California
Seismic Network

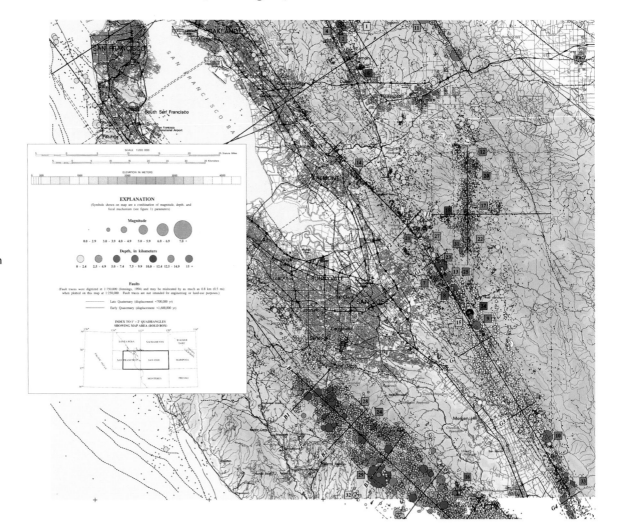

This map is the first in a series of seismicity maps that will cover the San Andreas Fault system and other seismically active areas in northern and central California. The Santa Rosa and Mariposa 1:250,000-scale quadrangles are currently in preparation. The work is supported by the Western Earthquake Hazards team of the U.S. Geological Survey (USGS), which produced maps to visually represent the information from the 32,000 earthquakes located by USGS in the San Francisco/San Jose region from 1967–1993.

Color is used to group earthquakes of the same mainshock/aftershock sequence and to distinguish them from other sequences and from the background activity. Lighter shades of color show the decay of the earthquake sequences over time. Earthquake cross sections created along and across major faults in many cases clearly delineate the dimensions and orientation of the faults in the upper 15 kilometers of the crust.

Epicenters are plotted on a colored relief base along with faults and cultural features. Symbol size and color indicate earthquake depths and magnitudes. Focal mechanisms overlaid on larger events show that dominant fault movement is right-lateral, strike-slip along northwest-oriented vertical faults.

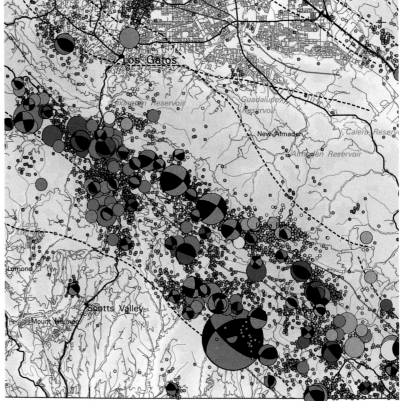

122° 00′

Geological Map of Iceland, 1:1,000,000

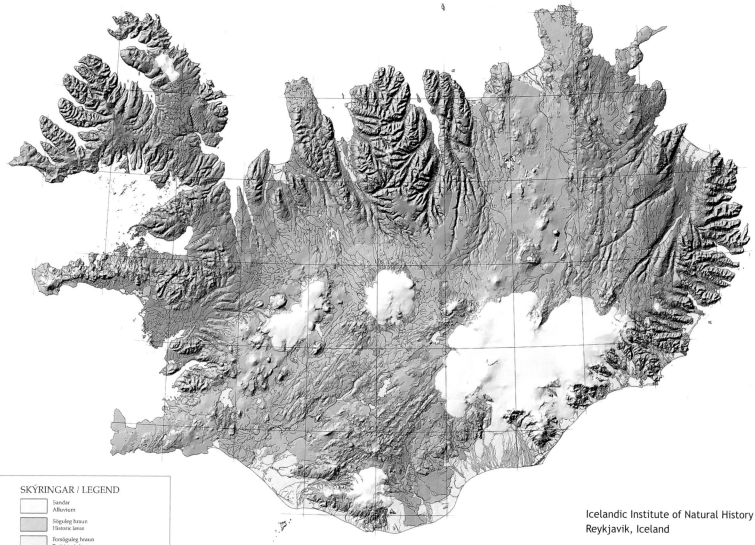

SKÝRINGAR / LEGEND

- Sandar
  Alluvium
- Söguleg hraun
  Historic lavas
- Forsöguleg hraun
  Prehistoric lavas
- Líparít
  Rhyolite
- Móberg frá síðari hluta ísaldar
  Late Ice Age hyaloclastites
- Hraun frá síðari hluta ísaldar
  Late Ice Age lavas
- Berglög frá síðtertíer og fyrri hluta ísaldar
  Late Tertiary and Early Ice Age bedrock
- Berglög frá tertíer
  Tertiary bedrock
- Djúpbergsinnskot
  Gabbro or granophyre

The geological map of Iceland shows the main features of the bedrock geology. Formations are classified by age, type, and composition. The map also depicts the volcanic zones of the island and the distribution of the recent eruption sites. Lava fields of the Holocene are shown as prehistoric or historic.

This map is based on data from the 1998 Geological Map 1:500,000, which was one of three natural history maps that won first place for Best Cartographic Publication and Best Overall Map Gallery Presentation at the 1998 ESRI International User Conference. Here, the bedrock geology is fused into a hill shade to create a painted relief, which gives a different perspective on the geology and the landscape.

Icelandic Institute of Natural History
Reykjavik, Iceland

*By Hans H. Hansen,
Haukur Jóhannesson, and
Kristján Sæmundsson*

Contact
Hans Hansen
hans@ni.is

Software
ArcInfo 7.1.2
Hardware
Sun SPARCstation
Printer
Professionally printed
Data Source(s)
Icelandic Institute of Natural History
1998 Geological Map 1:500,000,
bedrock geology

Geologic Map Atlas of the Villa Grove Quadrangle, Illinois: Coal Resource Map

Illinois State Geological Survey
Champaign, Illinois

By Pam Carrillo and Colin Treworgy

Contact
Colin Treworgy
colin@isgs.uiuc.edu

Software
ArcInfo 7.2.1, CorelDraw,
EarthVision 5.0, and GNU Ghostscript

Hardware
Sun SPARCstation, PCs with
Windows NT, and SGI

Printer
HP DesignJet 750C+

Data Source(s)
In-house, U.S. Geological Survey
digital raster graphic for base map

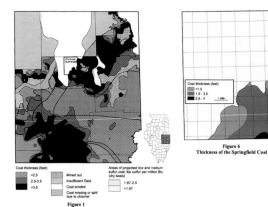

Figure 1
Thickness of the Herrin Coal in east-central Illinois

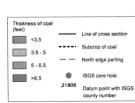

Figure 6
Thickness of the Springfield Coal

Figure 7a
Depth to top of the Herrin Coal (main bench)

Figure 7b
Thickness of the bedrock above the Herrin Coal (main bench)

Figure 7   The thickness and composition of earth materials overlying the Herrin Coal. The Springfield Coal lies 15 to 25 feet below.

Figure 7c
Thickness of the drift above the Herrin Coal

Figure 7d
Ratio of the thickness of bedrock to drift above the Herrin Coal (main bench)

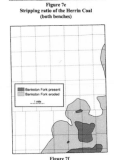

Figure 7e
Stripping ratio of the Herrin Coal (both benches)

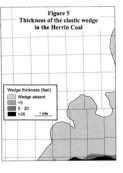

Figure 7f
Areas of Herrin Coal overlain by Bankston Fork Limestone

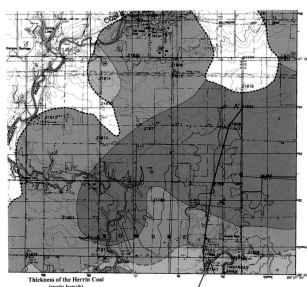

Thickness of the Herrin Coal
(main bench)

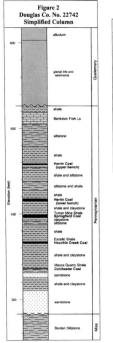

Figure 2
Douglas Co. No. 22742
Simplified Column

This map shows the coal resources and availability for mining for the Villa Grove Quadrangle, Douglas County, Illinois. It is one of a set from an atlas of detailed geologic information and derivative maps.

Interpreted geologic maps and derivative maps, such as the coal resources map, are important tools for implementing "smart growth." Community officials and urban planners can make better land use decisions regarding natural resources when they use these informative maps. Coal mining companies also use maps such as this one to guide their investment and mine planning decisions.

Figure 4
Thickness of the Herrin Coal
(upper bench)

Figure 5
Thickness of the clastic wedge
in the Herrin Coal

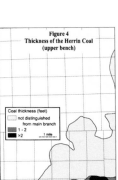

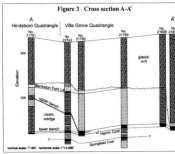

Figure 3   Cross section A-A'

Mapping Vertical Relief Dissection Using GIS

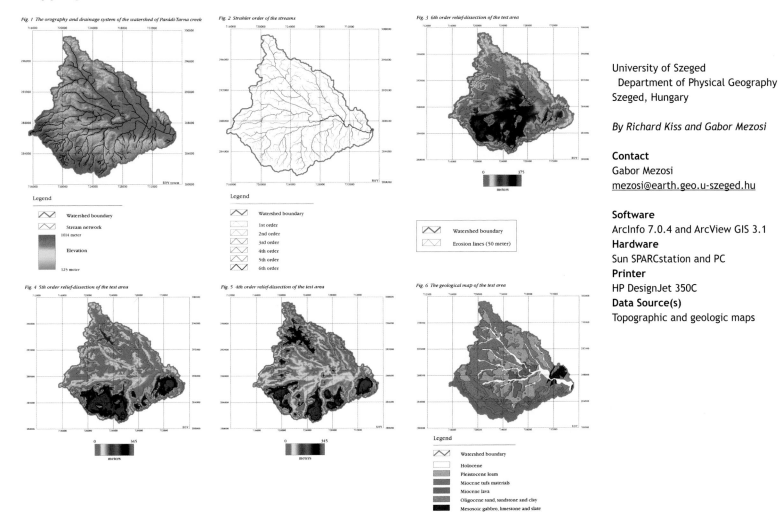

Fig. 1 The orography and drainage system of the watershed of Parádi-Tarna creek

Legend
- Watershed boundary
- Stream network
- 1014 meter
- Elevation
- 125 meter

Fig. 2 Strahler order of the streams

Legend
- Watershed boundary
- 1st order
- 2nd order
- 3rd order
- 4th order
- 5th order
- 6th order

Fig. 3 6th order relief-dissection of the test area

Fig. 4 5th order relief-dissection of the test area

Fig. 5 4th order relief-dissection of the test area

Fig. 6 The geological map of the test area

Legend
- Watershed boundary
- Erosion lines (50 meter)

Legend
- Watershed boundary
- Holocene
- Pleistocene loam
- Miocene tufa materials
- Miocene lava
- Oligocene sand, sandstone and clay
- Mesosoic gabbro, limestone and slate

University of Szeged
  Department of Physical Geography
Szeged, Hungary

By Richard Kiss and Gabor Mezosi

Contact
Gabor Mezosi
mezosi@earth.geo.u-szeged.hu

Software
ArcInfo 7.0.4 and ArcView GIS 3.1
Hardware
Sun SPARCstation and PC
Printer
HP DesignJet 350C
Data Source(s)
Topographic and geologic maps

Studying the erosion development of a given surface is a traditional subject in geomorphology. In this project, the authors attempted to find out the limits of GIS in morphometrical analysis. The aim was not only to produce and analyze a vertical relief dissection map, but also to look at the regional difference in erosion and its translocation in a geological scale.

A vertical relief dissection map could be constructed for each stream order by subtracting the real surface from the summit planes fitting on the watersheds. Such a map can give information about long-term translocations of erosion and about the changes in its rate. Mapping of vertical relief dissection based on morphometrical analyses was attempted in the 1970s, but it had limitations because of the lack of suitable methods. Surface modeling with GIS provided the right tool.

The analyses were done on an approximately 100-square-kilometer catchment area. It is situated on the northeast edge of the 15-million-year-old Mátra volcano, which produced mostly andesite and riolithe. The test area was more or less geologically homogeneous, and the possibility of translocation of stream directions could be accounted for. Additionally, precipitation conditions are very similar and did not significantly change the rate of erosion.

Stream order is a classification of streams based on tributary junctions and has proven to be a useful indicator of stream size, discharge, and drainage area. On a topographic map showing all intermittent and perennial streams in a basin, the smallest unbranched tributaries are designated order 1. Where two first-order streams join, a second-order stream segment is formed; where two second-order segments join, a third-order segment is formed, and so on.

The authors assumed that today a six-order stream was previously five-order, four-order even earlier, and so on. This is true only as a model, because despite the geological homogeneity, climatological and orographical disturbances could occur. The aim of the research was to focus on the disturbances in the drainage network.

There is an exponential relationship between the stream orders and the lengths of certain stream orders, and the map shows that the number of the five-order streams (as well as the six-order ones) is smaller than expected. Their mean lengths and volumes belong to the vertical relief dissections and are decreasing radically compared to statistical exceptions. It can be explained partly lithologically—the five-order rivers are situated in the middle of the catchment, in an erosion resistant environment—and partly by decreasing relative relief.

Parachilna 1:250,000 Geological Map

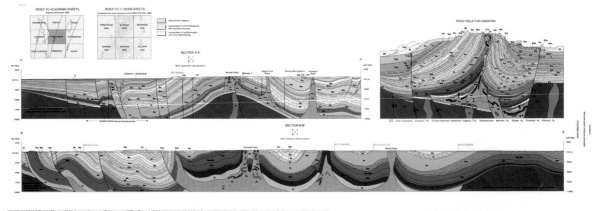

Primary Industries and Resources South Australia
Adelaide, South Australia

By John Bradford and Peter Reid

Contact
Sergio Rossi
rossi.sergio@saugov.sa.gov.au

Software
ArcInfo 7.2.1
Hardware
Sun UNIX workstations
Printer
Five-color Heidelberg Speedmaster offset printing press
Data Source(s)
Primary Industries and Resources South Australia and student research

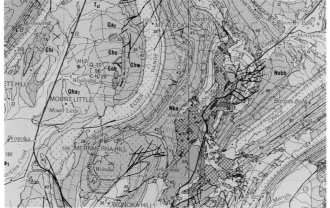

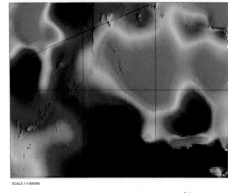

TOTAL MAGNETIC INTENSITY IMAGE

The Parachilna geological map, along with others in the 1:250,000 Geological Atlas of South Australia, depicts and partly interprets the geology of a discrete region of the state. The map includes a geological reference, cross sections, rock-relation diagrams, and other geoscientific information. It was compiled by the geological staff of Primary Industries and Resources South Australia (PIRSA) from various sources including departmental mapping, company mineral exploration, and investigation by university students.

The main purpose of the map is to promote responsible mineral exploration through a better understanding of geology. However, as the map area covers some of the most scenic parts of the Flinders Ranges, bushwalkers and tourists who want to learn about the rocks they are passing over also regularly use the map.

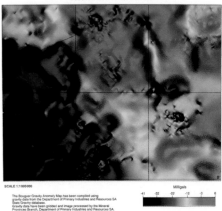

BOUGUER GRAVITY ANOMALY MAP

Cartographers in the PIRSA Spatial Information Services Branch carried out all stages of map preparation and used cartographic options from ArcInfo to achieve the high standard required for publication. The layout was assembled using ARC Macro Language (AML) to display numerous ArcInfo coverages from the South Australia geology database.

ArcInfo graphic files were generated and converted into digital color separation files using the ARC separator option. The separation files were then used to print negative films on a high-resolution image setter for making the printing plates. The map was printed in five colors on Impress Matt 115 gsm paper using a five-color Heidelberg Speedmaster offset printing press.

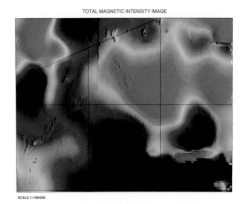

TOTAL MAGNETIC INTENSITY IMAGE

A further benefit of this process is the availability of digital geological data for the map sheet area. This is not a hard-copy plot of near-publication quality but an actual printed product (1,500 copies). This is the fourth map in the 1:250,000-scale series produced by this method, and adoption of the technology has resulted in a significant reduction in production time compared to traditional cartographic methods.

Geology of Tasmania

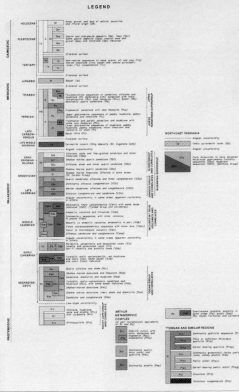

Mineral Resources Tasmania
Hobart, Tasmania

*By A.V. Brown, C.R. Calver,
M.J. Clarke, K.D. Corbett,
J.L. Everard, S.M. Forsyth,
B.A. Goscombe, G.R. Green,
M.P. McClenaghan, J. Pemberton,
and D.B. Seymour*

Contact
Ken Bird
kbird@mrt.tas.gov.au

Software
ArcInfo 7.2.1
Hardware
Sun Ultra 2 workstation
Printer
HP DesignJet 2500CP
Data Source(s)
Tasmanian Geological Survey
Geological Atlas 1:250,000

This map depicts the geology of Tasmania at a scale of 1:500,000. It is widely used by the government, public, and mining industry.

Geological field mapping in Tasmania is carried out using a 1:25,000-scale topographical base. Digital data and hard-copy output also are produced at this scale. The 1:25,000-scale geology data is manually simplified for both digital and hard-copy output at 1:250,000.

To produce the data from which the map is derived, ArcInfo was used to further generalize the 1:250,000-scale digital geology data to produce this map. This generalization process uses ARCPLOT commands in ARC Macro Language (AML) to erase, amalgamate, and simplify the 1:250,000-scale data for output at the smaller scale.

Using this methodology, a "live relationship" is maintained across the three scales of data used for geological map output in Tasmania.

Technical Cooperation with Chile, Environmental Geology for Regional Planning: Environmental Geological Map for the Area of Puerto Montt, Chile, 1:100,000

Federal Institute for Geosciences and
  Natural Resources (BGR)
Hannover, Germany

*By Dagmar Moellers and Markus
Toloczyki*

Contact
Dagmar Moellers
dagmar.moellers@bgr.de

Software
PC ARC/INFO 3.4.2, ArcView GIS 3,
and Microsoft Access
Hardware
PC586 and Pentium I
Printer
HP DesignJet 650C and 1055CM
Data Source(s)
Geological Survey of Chile, Santiago;
and BGR implementation of the
Department for Environmental
Geology

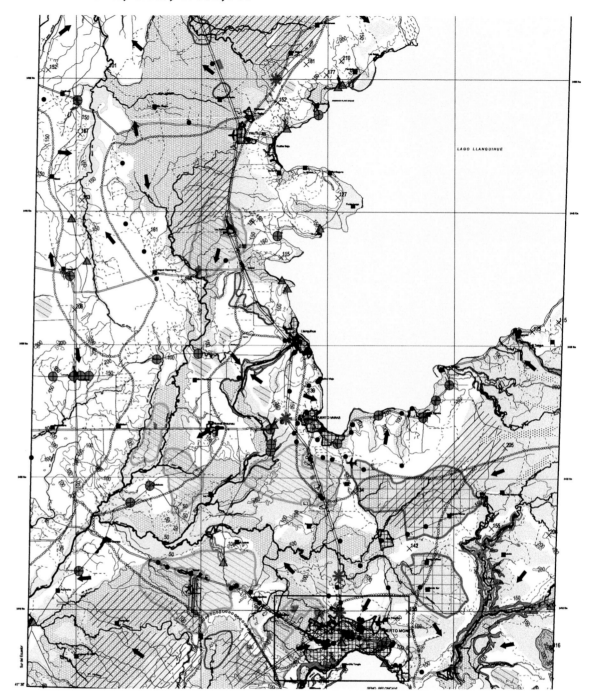

This map is the result of a cooperative effort between the Geological Survey of Chile and the Federal Institute for Geosciences and Natural Resources in Hannover, Germany.

The Environmental Geological map in 1:100,000 scale for the area of Puerto Montt was compiled from data collected in 1999. The project area is located in the central depression between the Coast Cordilleras and the Andes Mountains. Lakes and hilly country suggest Pleistocene glaciation. The altitude extends from sea level to about 180 meters above sea level. The climate is temperate, cold, and humid. Puerto Montt is a provincial capital with a population of 200,000.

The map covers the following topics: (1) protection of groundwater resources, (2) conservation of near surface mineral resources, (3) identification of potential areas for waste disposal sites, and (4) warnings for natural risks and hazards.

Thematic maps were produced and integrated into the Synthesis map. Themes included are geology, aquifers, urban development, natural and geological hazards, groundwater resources, and mineral resources. Planners use this map to quickly identify areas of competing land uses or locate areas at risk for disasters.

GIS in Missouri City, Texas

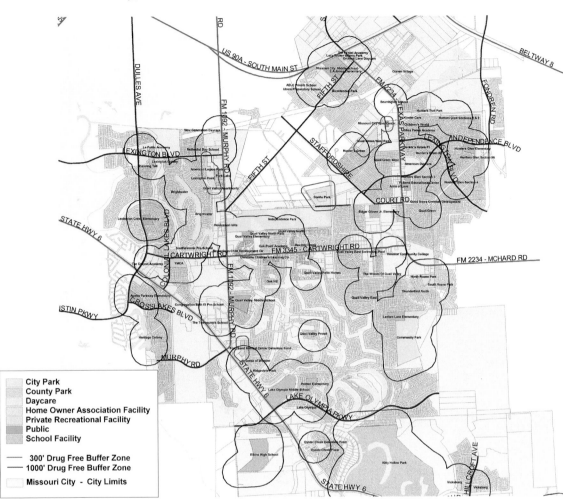

Legend

City Park
County Park
Daycare
Home Owner Association Facility
Private Recreational Facility
Public
School Facility

——— 300' Drug Free Buffer Zone
——— 1000' Drug Free Buffer Zone
Missouri City - City Limits

Public Works/Engineering Division
City of Missouri City, Texas

By Mark Hochstein

Contact
Mark Hochstein
markh@ci.mocity.tx.us

Software
ArcInfo 7.2.1 and ArcView GIS 3.1
Hardware
IBM Pentium II
Printer
HP DesignJet 650C
Data Source(s)
Missouri City GIS and tax data,
Fort Bend County Central Appraisal
District, state of Texas licensed day
care database

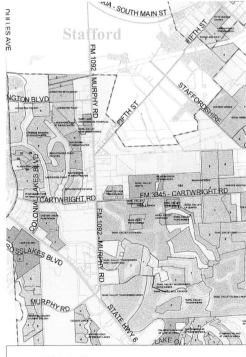

MISSOURI CITY
MISSOURI CITY EXTRA TERRITORIAL JURISDICTION
MISSOURI CITY - CITY LIMITS

The Residential Subdivision map was created from Missouri City's parcel coverage and tax database. Using ArcInfo, subdivision numbers were selected within the parcel coverage and then dissolved to create the new coverage. Minor cleanup was performed to finalize the coverage. All city departments use this map.

The Missouri City Police Department is able to put more "bite" into the enforcement of local drug laws with the Drug Free Zone map. The buffered areas on the map, created in ArcView GIS, are zones where drug violations are doubled in penalty. This helps eliminate drug violators getting off with light sentences.

The idea to create this map came from a neighboring city's police and engineering departments. The city "manually" created a similar type of map, which used up man-hours of several employees over a period of one month. In less than eight man-hours, one Missouri City employee using ArcView GIS, existing GIS coverages, and free state of Texas data acquired via the Internet created an improved map.

GIS Activities in Support of Indian Country

Bureau of Indian Affairs
 Geographic Data Service Center
Lakewood, Colorado

By Dennis Marenger

Contact
Dennis Marenger
dmarenger@gdsc.bia.gov

Software
ArcInfo 7.2.1, ARC GRID
Hardware
Compaq computers and Sun 690
servers
Printer
HP DesignJet 755CM
Data Source(s)
Bureau of Indian Affairs, U.S.
Geological Survey, and U.S. Census
Bureau

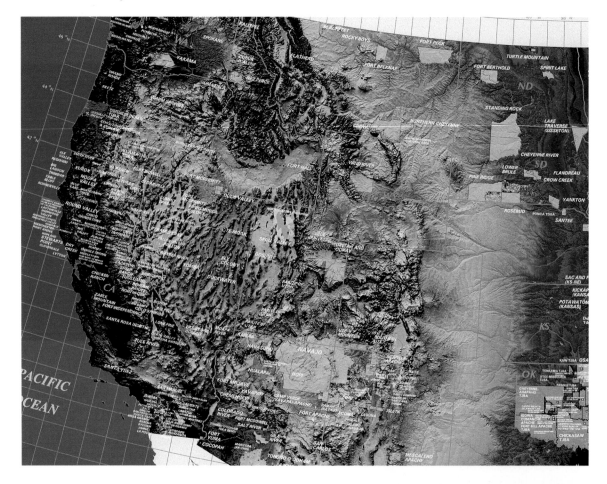

For 12 years the Bureau of Indian Affairs Geographic Data Service Center (GDSC) has been developing special map products in support of tribal and Bureau of Indian Affairs offices. The GDSC has developed GIS databases, which enable the production of map products. Some maps are general, used mainly for overview and regional-scale analysis, while others are specific and address important issues in Indian country. Indian Lands in the United States and Native Entities within the State of Alaska represent examples of small-scale maps used mainly for general overview.

The Indian Lands in the United States map was produced to replace an older map developed in the 1960s. It depicts the current geographic distribution of Indian entities recognized and eligible to receive services from the Bureau of Indian Affairs as published in the *Federal Register,* October 1997. The map represents some 55.7 million acres of land held in trust by the United States and locates more than 300 tribal entities in the contiguous United States. This map also shows tribal entities recognized by individual states. The shaded relief on the map was generated with ARC GRID software and represents 250 elevation samples at 50-foot intervals. This map product is available from the U.S. Geological Survey at 1-888-ASK-USGS.

The same map concept was applied to the Native Entities within the State of Alaska map. In Alaska, there is only one federally recognized American Indian reservation—Annette Islands Reserve. More than 200 native entities have properties that are managed by one of the 13 Alaska Native Regional Corporations (ANRCs). This map product is also available for sale from the U.S. Geological Survey.

The Indian country maps won the Best Data Integration award at ESRI's 1999 User Conference Map Gallery display.

South African Development Community, Demographic, Social, and Economic Characteristics

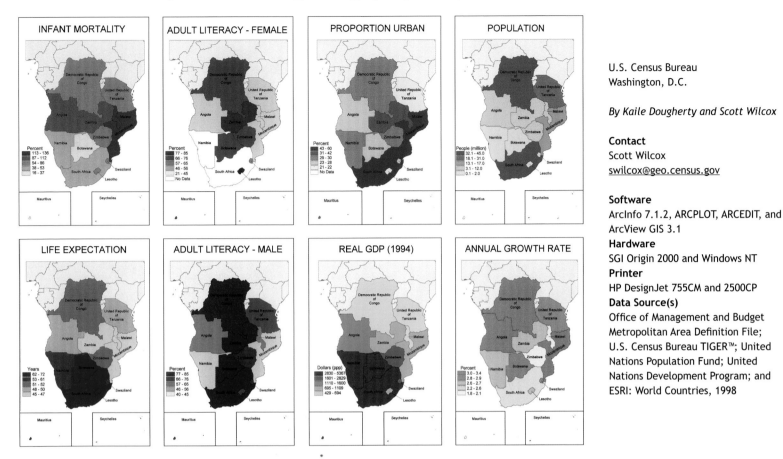

INFANT MORTALITY

Percent
113 - 136
87 - 112
54 - 86
38 - 53
16 - 37

ADULT LITERACY - FEMALE

Percent
77 - 85
66 - 76
57 - 65
46 - 56
21 - 45
No Data

PROPORTION URBAN

Percent
43 - 60
31 - 42
29 - 30
23 - 28
21 - 22
No Data

POPULATION

People (million)
32.1 - 45.0
18.1 - 31.0
13.1 - 17.0
3.1 - 12.0
0.1 - 2.0

LIFE EXPECTATION

Years
62 - 72
53 - 61
51 - 52
48 - 50
45 - 47

ADULT LITERACY - MALE

Percent
77 - 85
66 - 76
57 - 65
46 - 56
40 - 45

REAL GDP (1994)

Dollars (ppp)
2830 - 5367
1601 - 2829
1110 - 1600
695 - 1109
429 - 694

ANNUAL GROWTH RATE

Percent
3.0 - 3.4
2.8 - 2.9
2.6 - 2.7
2.2 - 2.6
1.8 - 2.1

U.S. Census Bureau
Washington, D.C.

By Kaile Dougherty and Scott Wilcox

Contact
Scott Wilcox
swilcox@geo.census.gov

Software
ArcInfo 7.1.2, ARCPLOT, ARCEDIT, and
ArcView GIS 3.1
Hardware
SGI Origin 2000 and Windows NT
Printer
HP DesignJet 755CM and 2500CP
Data Source(s)
Office of Management and Budget
Metropolitan Area Definition File;
U.S. Census Bureau TIGER™; United
Nations Population Fund; United
Nations Development Program; and
ESRI: World Countries, 1998

The South African map was created in support of a workshop on census mapping held in South Africa in 1999. It shows a comprehensive view of each of the participating countries. The map demonstrates that although the countries are relatively close, their respective characteristics are quite dissimilar.

Metropolitan Areas of the United States
The Metropolitan Area map displays the geographic extent and names of the four different types of metropolitan areas as defined by the Office of Management and Budget for the United States and Puerto Rico.

The four types are Consolidated Metropolitan Statistical Areas (CMSAs), Metropolitan Statistical Areas (MSAs), Primary Metropolitan Statistical Areas (PMSAs), and New England County Metropolitan Areas (NECMAs). The map content includes population as determined from the 1990 decennial census, Census Bureau-defined urbanized area extent, and the component county or equivalent name for each metropolitan area. ArcInfo was used to develop cartographic boundary files from the Census Bureau TIGER database. Annotation layers for each of the boundary layer types were prepared interactively. An ARCPLOT ARC Macro Language file generated the map image.

There is more information about Census Bureau map products at this Web site: www.census.gov.

Metropolitan Areas of the United States

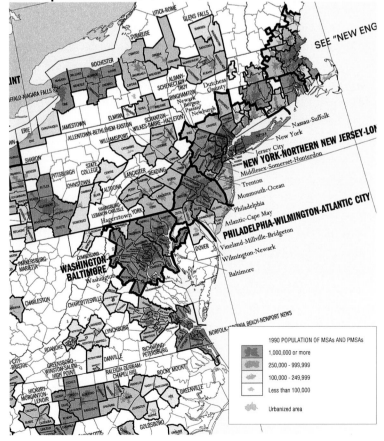

1990 POPULATION OF MSAs AND PMSAs
1,000,000 or more
250,000 - 999,999
100,000 - 249,999
Less than 100,000
Urbanized area

Cape Cod Breast Cancer Study

Silent Spring Institute
Newton, Massachusetts
Applied Geographics, Inc.
Boston, Massachusetts

By Yvette Joyce and Steven Melly

Contact
Steven Melly
melly@silentspring.org

Software
ArcView GIS 3
Hardware
Sun SPARC 20 workstation
Printer
HP DesignJet 2500CP
Data Source(s)
Silent Spring Institute, Massachusetts
Cancer Registry, Massachusetts
Executive Office of Environmental
Affairs, MassGIS, Massachusetts
Department of Environmental
Management, Cape Cod Commission,
University of Massachusetts Resource
Mapping Project, and U.S. Census
Bureau

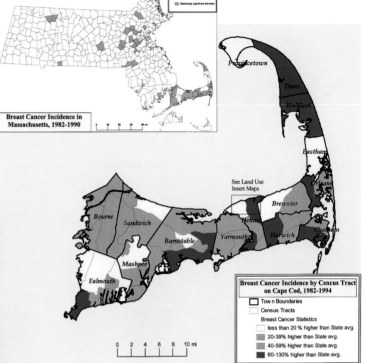

Breast Cancer Incidence in Massachusetts, 1982-1990

Breast Cancer Incidence by Cencus Tract on Cape Cod, 1982-1994

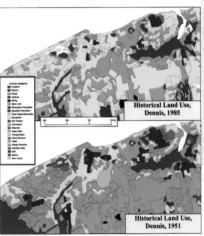

Historical Land Use, Dennis, 1985

Historical Land Use, Dennis, 1951

60 | health care

GIS enabled the study of the incidence of breast cancer at the town, tract, and census block group level, which was part of the Silent Spring Institute's work studying breast cancer and the environment on Cape Cod.

Using parcel data, the addresses of 2,625 women diagnosed with breast cancer from 1982 to 1994 were geocoded. Land use data helped to refine population estimates by age group. These population estimates were used to determine the number of breast cancer cases expected in an area based on the rate of breast cancer in the rest of the state.

The standardized incidence ratios (shown on the map) were calculated in order to see how breast cancer incidence in the census tracts of the cape compared to the rest of the state while taking into account differences in the age distribution of the population. This map shows that breast cancer incidence varies within the cape, but areas of elevated incidence are scattered throughout the area. The study has now shifted to a focus on regional environmental factors as the search for clues to understanding the pattern of elevated incidence continues.

GIS is now being used to assess environmental exposures to 2,500 women with and without breast cancer on Cape Cod. Women are interviewed to find out where they lived on the cape during the past 40 years and to discover information about individual risk factors and environmental and occupational exposures.

The GIS provides information about potential sources of exposure such as areas where pesticides were applied. The women themselves might not be aware of this data. GIS helps to consider multiple factors such as attributes of the source, distance and direction from the source to the subject's residence, and whether there are intervening factors such as a forested area that might reduce a subject's exposure to a source.

Silent Spring's GIS includes land use data based on aerial photography from four points in 1951, 1971, 1984, and 1990. Maps such as the one shown here enable researchers to identify environmental features that might be associated with health effects. Environmental features considered in the study of breast cancer include pesticide use areas such as agriculture, cranberry bogs, and golf courses. The historical data helps to identify locations that are now residential but where pesticides were used in the past. Land use maps also help to identify areas where groundwater used for drinking might be impacted by septic tanks associated with residential and commercial development.

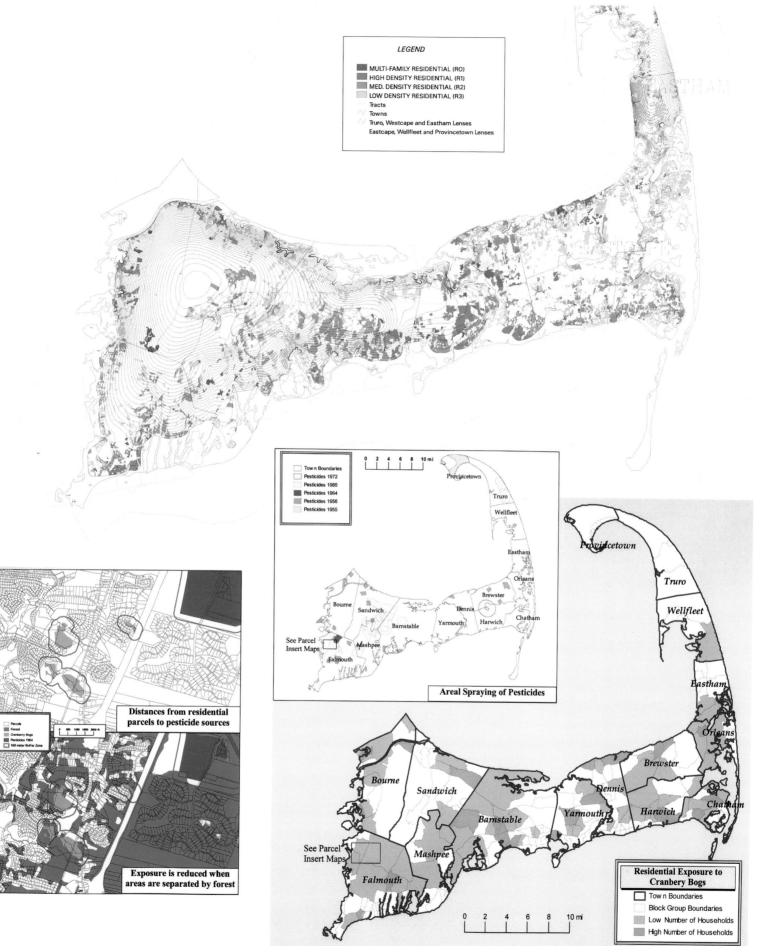

LEGEND

MULTI-FAMILY RESIDENTIAL (RO)
HIGH DENSITY RESIDENTIAL (R1)
MED. DENSITY RESIDENTIAL (R2)
LOW DENSITY RESIDENTIAL (R3)
Tracts
Towns
Truro, Westcape and Eastham Lenses
Eastcape, Wellfleet and Provincetown Lenses

Town Boundaries
Pesticides 1972
Pesticides 1965
Pesticides 1964
Pesticides 1956
Pesticides 1955

0 2 4 6 8 10 mi

Provincetown
Truro
Wellfleet
Eastham
Orleans
Brewster
Bourne
Sandwich
Dennis
Chatham
Barnstable
Yarmouth
Harwich
See Parcel
Insert Maps
Mashpee
Falmouth

Areal Spraying of Pesticides

Parcels
Forest
Cranberry Bogs
Pesticides 1964
100 meter Buffer Zone

0 500 1000 1500 2000 ft

Distances from residential parcels to pesticide sources

Exposure is reduced when areas are separated by forest

Provincetown
Truro
Wellfleet
Eastham
Orleans
Bourne
Sandwich
Brewster
Dennis
Barnstable
Yarmouth
Harwich
Chatham
Mashpee
See Parcel
Insert Maps
Falmouth

Residential Exposure to Cranberry Bogs

Town Boundaries
Block Group Boundaries
Low Number of Households
High Number of Households

0 2 4 6 8 10 mi

...ng the Value of Location: Improving Estimations Through GIS and Spatial Autocorrelation ...iques

City of Regina
Regina, Saskatchewan

By Roberto Figueroa

Contact
Roberto Figueroa
rfiguero@cityregina.com

Software
ArcInfo 7.1.3, ArcView GIS 3.1, and
SPSS Version 8.0
Hardware
IBM RS6000 F50 and IBM PC 300PL
Printer
HP DesignJet 2500CP
Data Source(s)
Property assessment database and
land title certificates

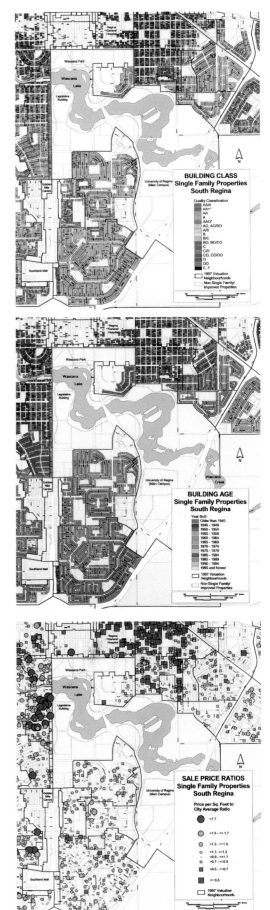

LOCATION VALUES
Single Family Properties
City of Regina

Premium Values ($)
- \>30000
- 20001 - 30000
- 10001 - 20000
- 0 - 10000

Penalty Values ($)
- -10000 - 0
- -20000 - -10001
- -30000 - -20001

BUILDING CLASS
Single Family Properties
South Regina

BUILDING AGE
Single Family Properties
South Regina

SALE PRICE RATIOS
Single Family Properties
South Regina

With a population of 195,000, the city of Regina is Saskatchewan's provincial capital and contains nearly 58,000 properties within 29,000 acres. Since 1997, the city's assessment division has been conducting property reassessments following new provincial standards and procedures dictated by the Saskatchewan Assessment Management Agency. The new methodologies are more market sensitive as they adjust cost schedules according to market adjustment factors (MAFs), which are based on prior comparable property sales.

The maps were designed to help visualize the city of Regina's property spatial data. Thematic maps include building age, lot size, building class/grade, and sale price ratios. In addition, the maps display the results of ongoing research focusing on the estimation of property location values in different geographic areas of Regina. The Location Values map depicts city areas that are receiving location premiums or penalties in property values because of their geographic location. The ability of GIS to link spatial and computer-assisted mass appraisal data has enabled the production of thematic maps that display property attributes, which support neighborhood amalgamations and/or realignments for the next property reassessment in 2001. Revision and analysis of location values are required to support decisions on changing neighborhood valuation boundaries.

GIS is an excellent tool not only for visualizing the geographic distribution of location values but also for modeling estimated property location values according to surrounding location residuals. GIS can calculate spatial autocorrelation statistics, which are valuable when testing the spatial consistency of location premiums and penalties. Research results show a clearer spatial picture of building characteristics and location values, which has certainly helped in the reconfiguration of valuation neighborhoods for the 2001 property reassessment.

GIS Applications in the City of Lakewood

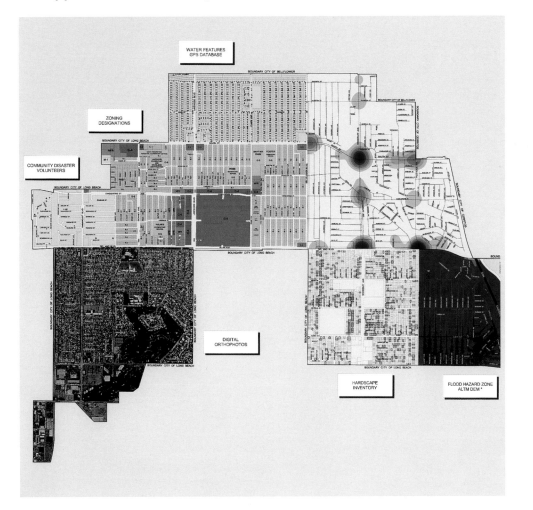

City of Lakewood
Lakewood, California

By Michael Jenkins

Contact
Sonia Southwell, GIS Coordinator
ssouthwe@lakewoodcity.org

Software
ArcInfo 7.2.1, ArcView GIS 3.1, ArcView
Spatial Analyst 1.1, ArcPress 2, and
Trimble Pathfinder Office 2.11
Hardware
HP 9000/735 series workstation,
Gateway Pentium II PC, Trimble
Pro XR GPS receiver, Trimble TSC1
Asset Surveyor, and Laser Atlanta
Advantage laser range finder
Printer
HP DesignJet 650C
Data Source(s)
Airborne Laser Terrain Mapping
(ALTM) data; digital elevation model
by Optech, Inc., Toronto; and
Dr. Roberto Gutierrez, University of
Texas at Austin

Since 1980, students at Lindstrom School are given a Suggested Route to School map at the beginning of each school year. The purpose of these maps is to provide students with a suggested route to and from school based on traffic volume, speed limits, visibility, and the location of traffic lights, stop signs, and crosswalks in their neighborhood.

Originally, the Suggested Route to School maps were developed and maintained manually. In 1999, all maps in the series were converted into a digital format using ArcView GIS software with funding provided by the California Office of Traffic Safety. GIS enables staff to update and reproduce the maps easily and frequently.

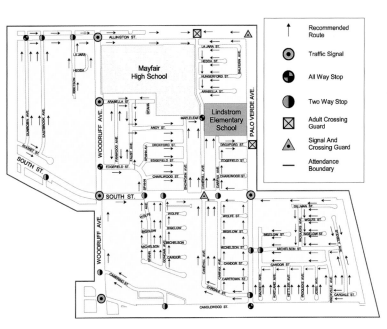

Base Map Series

Applied Geographics, Inc. (AGI)
Boston, Massachusetts

By David Bailey, Michael Terner, and Nick Wilkoff

Contact
Michael Terner
mgt@appgeo.com

Software
ArcView GIS 3.1
Hardware
350 MHz Windows NT PC
Printer
HP DesignJet 650C
Data Source(s)
Scanned image of original assessor's sheet, 100-scale digital planimetric data from EarthData International, and original vector parcel line work

(opposite top left)
County of Henrico, Virginia
Richmond, Virginia

By Alfredo Frauenfelder

Contact
Alfredo Frauenfelder
fra01@co.henrico.va.us

Software
ArcView GIS 3.1
Hardware
Windows PC
Printer
HP DesignJet 1055CM
Data Source(s)
County of Henrico, Virginia, digital ortho aerial photography and planimetric base map

Parcel Check Plot, Mashpee, Massachusetts—Sheet 105

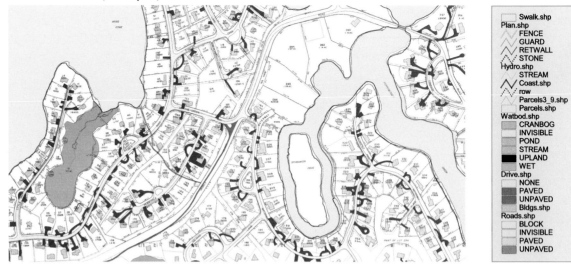

Parcel Check Plot, Mashpee, Massachusetts

Created as part of Mashpee's parcel automation project, this map shows the interrelationship between the town's original parcel maps and the town's new planimetric data created from a 1998 aerial flyover. The coastline, buildings, fences, road edges, and driveways are from the flyover data. This is a wholly working map with minimal map marginalia. The intent is to show the assessor the fit/discrepancies between the original parcel mapping and the new photogrammetric base map. The assessor's staff will mark up these maps to determine which parcel lines should be shifted to better match the new planimetry. The marked up check plots will then guide a second round of vector parcel line editing.

Digital Orthoimagery and Planimetric Base Map of Henrico County, Virginia

Henrico County, Virginia, has a GIS program that seeks to increase efficiency and the level of service delivery by providing state-of-the-art tools for the modern management of geographic information. Procurement of digital imagery and related products was the first step in establishing an accurate and precise geographic information database to support the functions throughout Henrico County. Subsequent efforts will provide the county with a digital parcel identification map and other information layers that comply with the county's GIS standards and specifications.

ArcCR City—GIS Database of the Czech Cities

ArcCR City Project is part of the data program of ARCDATA PRAHA, a database of spatial information about cities in the Czech Republic. The data attributes include address points, street lines, urban units, parcel units, basic land use, and raster town plans. This data is accessible to GIS users, and its contents and structure are especially suited for business applications such as geocoding, network analyses, and address location.

Technical Map of Ostrava

Ostrava is the third largest town in the Czech Republic. Managers in the field of energy distribution and public administration use the Technical Map of Ostrava as a planning tool. Aerial photogrammetry and overland surveying data were used to produce the map. It shows topography, buildings, roads, railways, parks, streams, and utility features such as street lighting and traffic signs.

Mapping New York City— Infrastructure to Environment

New York City government is responsible for providing hundreds of services such as garbage collection, pothole repair, catch basin cleaning, meter reading, tax assessment, fire and police protection, public transportation, and inspections. Because city services are predominantly location-driven, maps that depict the physical location and property boundaries have always been used as a tool to support government operations. In an effort to place its water main and sewer facilities on a common base map, the Department of Environmental Protection began creating an accurate physical base map of the city. The project, scheduled for completion early in 2000, will create a citywide GIS utility.

Digital Orthoimagery and Planimetric Base Map of Henrico County, Virginia

Technical Map of Ostrava

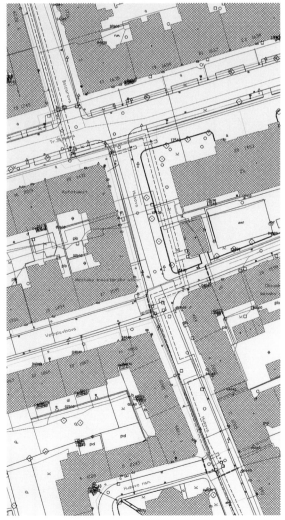

ArcCR City—GIS Database of the Czech Cities

Mapping New York City— Infrastructure to Environment

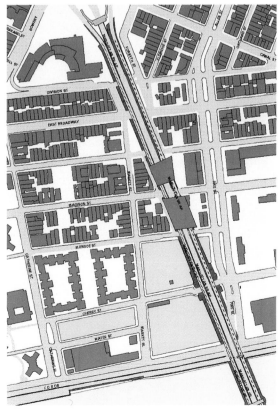

(top right)
ARCDATA PRAHA
Prague, Czech Republic

By Jan Vodnansky

Contact
Jan Vodnansky
vodnansky@arcdata.cz

Software
ArcView GIS 3.1
Hardware
DEC PW Pentium II
Printer
HP DesignJet 750C+
Data Source(s)
Land Survey Office, Institute for
Municipal Informatics of Capital
Prague, City of Ostrava, and P.F. Art

(bottom left)
DIGIS Ostrava Ltd.
Ostrava, Czech Republic

*By Pavel Bentkowski, Mojmir
Novacek, and Milada Novackova*

Contact
Libor Stefek
lstefek@digis.cz

Software
ArcCAD 11.4.1
Hardware
PC Pentium II
Printer
HP DesignJet 2500CP
Data Source(s)
Local government

(bottom right)
City of New York
New York, New York

*By Wendy Dorf, Paul Katzer, and
David LaShell*

Contact
Paul Katzer
pirate@parklan.cn.ci.nyc.ny.us

Software
ArcInfo 8, ArcInfo 7, ArcView GIS 3,
and ArcView Spatial Analyst 1.1
Hardware
Windows NT Intel

Seminole County, Florida—Integrating Local Government with GIS

Seminole County Government
 Information Technologies/GIS
 Division
Sanford, Florida

By Nancy Church and Albert Hill

Contact
Albert Hill
ahill@co.seminole.fl.us

Software
ArcInfo 7.1, ARCPLOT, and
ArcView GIS 3.1
Hardware
IBM RS6000 workstation
Printer
HP DesignJet 650C and 5000
Data Source(s)
Seminole County (A complete data
dictionary with more than 100 layers
of data is available online at
www.co.seminole.fl.us/gis.)

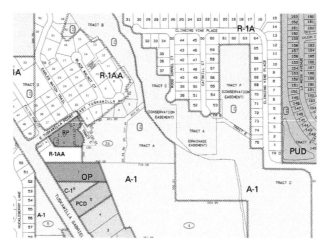

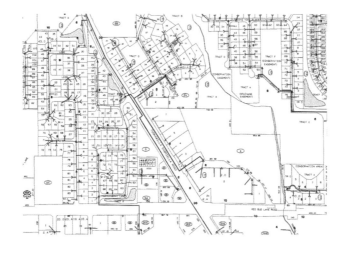

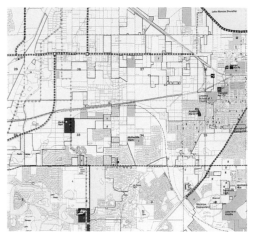

Since 1994, when the Board of County Commissions GIS program began, more than 150 layers
of data have been automated for county department and division use. Seminole County GIS takes
advantage of existing fiber wide area network and local area networks to tie together widely separated
county offices for efficient and effective data sharing. A central GIS server houses the county digital
map library.

Minutes north of Orlando, Florida, Seminole County offers plenty of the great outdoors including
18 major public parks, dozens of municipal parks, an extensive trails network, and more than 4,000
acres of nature preserves. Several county departments and agencies use the Trails and Greenways map
to manage and enhance these natural resources.

Safeguarding its citizens is one of the county's most important jobs. The Emergency Preparedness
map is a compilation of many layers that are important to this process. The layers are generated
and maintained by a number of internal agencies and plotted from a central map library to any of
a dozen plotters located countywide.

The county Emergency Operation Center dynamically displays the information depicted on this map
on large wall monitors, which facilitates communication and coordination during an activation.

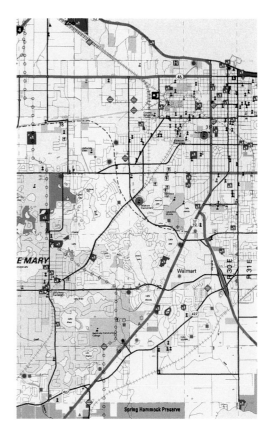

Integrating Geographic Information Science and Computer-Assisted Mass Appraisal

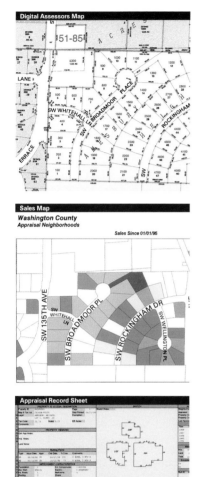

Washington County Cartography and
  the Engineering Support Unit
Salem, Oregon

*By Todd Davilla, Phillip DeMeyer,
Brian Ford, Jon Greninger, Roger
Livingston, Andrew Metz, Tim Spear,
John Swann, and Steven Wilson*

Contact
Jon Greninger
jon_greninger@co.washington.or.us

Software
ArcInfo 7.1.2, ArcView GIS 3.1, ARC
GRID, and Visual Basic Version 6.0
Hardware
Pentium II and PC
Printer
HP DesignJet 755CM
Data Source(s)
Recorded plats, surveys, and deeds

This cartographic product illustrates how the multidisciplinary field of GIS can aid in the process of computer-assisted mass appraisal. It is the result of an interagency partnership between Washington County, Oregon, and the Engineering Support Unit, a business division of Inside Oregon Enterprises.

In 1991, Washington County began to implement its GIS for a variety of planning-based applications. The main tax lot base map coverage used in county analysis and map production only represented a "planning level" base. While satisfactory for the majority of GIS applications, it had various degrees of accuracy and was not based on a tightly controlled geodetic network. Initially, the planning base was adequate for appraisal purposes.

In eight years, map production at Washington County has evolved into a more geographic information science-based process, integrating a variety of components known as geomatics (the study of cartography, surveying, and remote sensing). The GIS-based applications are particularly suited to help the appraisal staff in the process of mass appraisal. As COGO subdivisions, surveys, and deed information are assembled and joined into a new seamless base map, they are registered to a series of control points.

One of the objectives of Washington County's remapping program is to develop a tax lot base map with a .5-foot accuracy encompassing the urban growth boundary. With this increased positional accuracy, appraisers will have immediate access to attributes such as property acreage, lot dimensions, and positional coordinates. When this data is overlaid with aerial photography, appraisers can determine the location and orientation of improvements on a subject property. They can also analyze the data with other theme coverages in the process of property valuation.

Additional applications and products used by Washington County appraisal staff to aid in the mass appraisal of property include sales maps, school district maps, floodplain determination maps, viewshed analysis, appraisal neighborhood maps, soil classification maps, building footprints and attributes, slope, aspect, and digital terrain modeling.

Martin County Utilities System Map Book

LBFH, Inc.
Palm City, Florida

By Javier Cisneros, Dave Coleman,
Henry Mogilevich, and Bill Orazi

Contact
Dave Coleman
dave-c@lbfh.com

Software
ArcView GIS 3.1
Hardware
PC
Printer
HP DesignJet 755CM
Data Source(s)
Martin County GIS and Martin County
Utilities as-built drawings

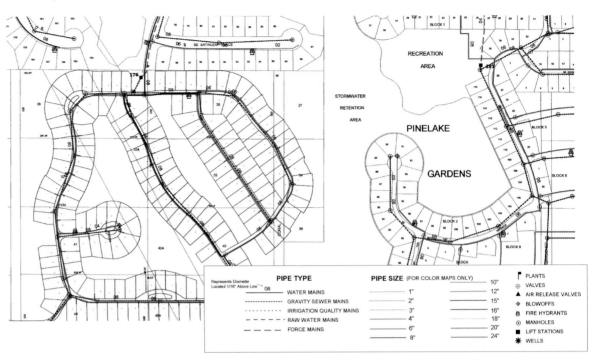

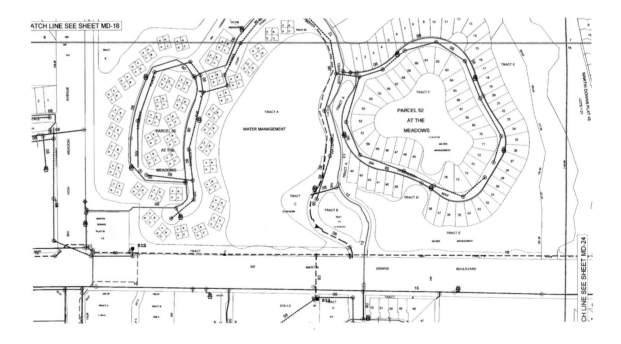

These images are part of a map book prepared by LBFH, Inc., and are a product of the GIS created for Martin County Utilities. The GIS augments tasks in operations, maintenance, planning, and customer service and uses various data sources and formats. The theme and attribute setup format were coordinated with Martin County Utilities for a customized graphical user interface. The resulting ArcView GIS project enabled technical and nontechnical staff to easily access and use the data.

The map book is particularly important to field personnel who do not have ready access to GIS. Office personnel also use the book, and a digital version of the book in PDF format enables yet another viewing option. During the next phase of the project, updates will be made such as adding new utility features, and a Web site is planned to make the data available to the public.

Zoning Map

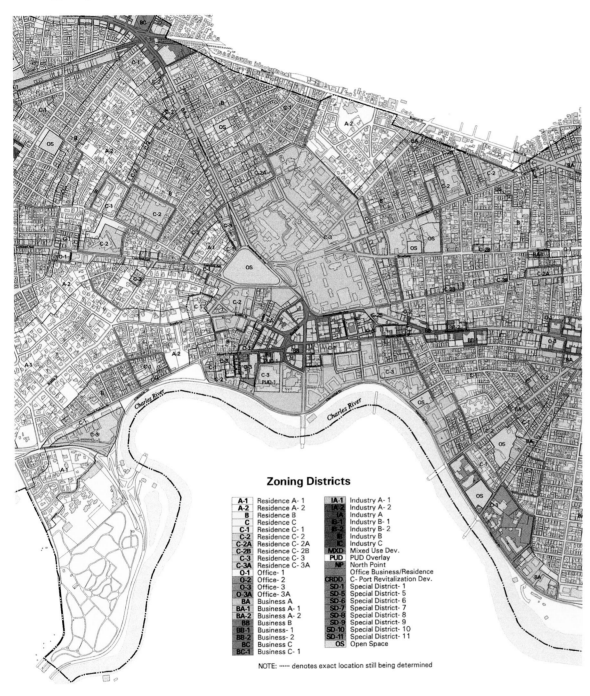

Zoning Districts

A-1	Residence A- 1		**IA-1**	Industry A- 1
A-2	Residence A- 2		**IA-2**	Industry A- 2
B	Residence B		**IA**	Industry A
C	Residence C		**IB-1**	Industry B- 1
C-1	Residence C- 1		**IB-2**	Industry B- 2
C-2	Residence C- 2		**IB**	Industry B
C-2A	Residence C- 2A		**IC**	Industry C
C-2B	Residence C- 2B		**MXD**	Mixed Use Dev.
C-3	Residence C- 3		**PUD**	PUD Overlay
C-3A	Residence C- 3A		**NP**	North Point
O-1	Office- 1			Office Business/Residence
O-2	Office- 2		**CRDD**	C- Port Revitalization Dev.
O-3	Office- 3		**SD-1**	Special District- 1
O-3A	Office- 3A		**SD-5**	Special District- 5
BA	Business A		**SD-6**	Special District- 6
BA-1	Business A- 1		**SD-7**	Special District- 7
BA-2	Business A- 2		**SD-8**	Special District- 8
BB	Business B		**SD-9**	Special District- 9
BB-1	Business- 1		**SD-10**	Special District- 10
BB-2	Business- 2		**SD-11**	Special District- 11
BC	Business C		**OS**	Open Space
BC-1	Business C- 1			

NOTE: ····· denotes exact location still being determined

City of Cambridge
Cambridge, Massachusetts

*By Roger Booth, Peter Bujwid,
Stuart Dash, Parvaneh Kossari, and
Robin Shore*

Contact
Peter W. Bujwid
pbujwid@ci.cambridge.ma.us

Software
ArcInfo 7.2.1
Hardware
UNIX (Alpha 4000)
Printer
HP DesignJet 755CM
Data Source(s)
City of Cambridge GIS

The city of Cambridge, with a population of 95,000 and a land area of 6.92 square miles, has more than 40 different zoning districts. Cambridge also has one of the most complex sets of zoning regulations in the country, which are continually revised to affect some of the districts. It is crucial to have an accurate and updated zoning map to monitor how zoning regulations affect property owners and neighborhood residents.

This map was developed in-house based on Community Development Department maps drawn in 1961. The individual zoning boundary lines were constructed by extracting parcel, road center, and city boundary lines from the Cambridge GIS database. The buffer command was also used to generate offset lines as needed.

This map is available in black and white or color at three different scales. You can view an online version of the zoning map at www.ci.cambridge.ma.us/~GIS/maplibrary/zoning.html.

Planning Maps

City of Temecula
Temecula, California

By Sharon Baker-Stewart, Riverside County
Kelli Beal and John De Gange, City of Temecula

Contact
Kelli Beal
bealk@co.riverside.ca.us

Software
ArcInfo 7.2.1
Hardware
HP Kayak Pentium II 333 and
Windows NT 400 workstation
Printer
HP DesignJet 2500CP
Data Source(s)
Riverside County parcel, centerline, and tax assessor data and City of Temecula zoning and vacant properties data

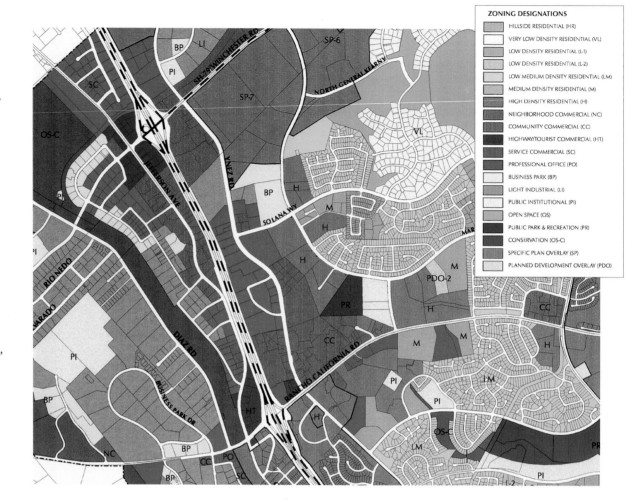

ZONING DESIGNATIONS

HILLSIDE RESIDENTIAL (HR)
VERY LOW DENSITY RESIDENTIAL (VL)
LOW DENSITY RESIDENTIAL (L-1)
LOW DENSITY RESIDENTIAL (L-2)
LOW MEDIUM DENSITY RESIDENTIAL (LM)
MEDIUM DENSITY RESIDENTIAL (M)
HIGH DENSITY RESIDENTIAL (H)
NEIGHBORHOOD COMMERCIAL (NC)
COMMUNITY COMMERCIAL (CC)
HIGHWAY/TOURIST COMMERCIAL (HT)
SERVICE COMMERCIAL (SC)
PROFESSIONAL OFFICE (PO)
BUSINESS PARK (BP)
LIGHT INDUSTRIAL (LI)
PUBLIC INSTITUTIONAL (PI)
OPEN SPACE (OS)
PUBLIC PARK & RECREATION (PR)
CONSERVATION (OS-C)
SPECIFIC PLAN OVERLAY (SP)
PLANNED DEVELOPMENT OVERLAY (PDO)

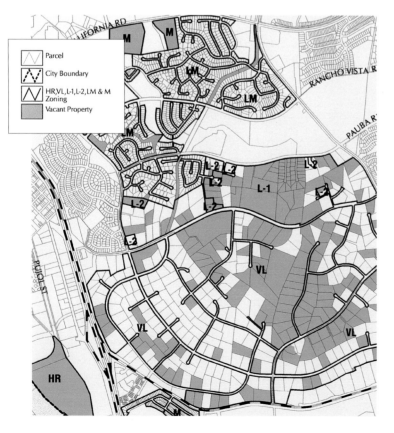

Parcel
City Boundary
HR, VL, L-1, L-2, LM & M Zoning
Vacant Property

Zoning maps serve to implement the land uses that will ultimately shape the lifestyles of the people who live, work, and shop in a community. When Temecula incorporated in 1989, the zoning that was in place had been established by Riverside County when the area was considered rural. Development and urbanization of the vacant land surrounding Temecula was ongoing and expansive in the late 1980s, and implementation of the county's zoning map was no longer suitable for this urban community. Accordingly, in 1995, Temecula adopted its zoning map, which accommodated the fast-paced growth the city was experiencing.

Large areas of vacant property remained within the city, and city officials wanted to analyze development patterns to effectively plan for the challenges of providing services and maintaining quality of life. The Vacant Properties map serves as a useful tool in these efforts. It also provides valuable information for the city's economic development program by easily pinpointing potential areas for development.

Denton County Soil Survey

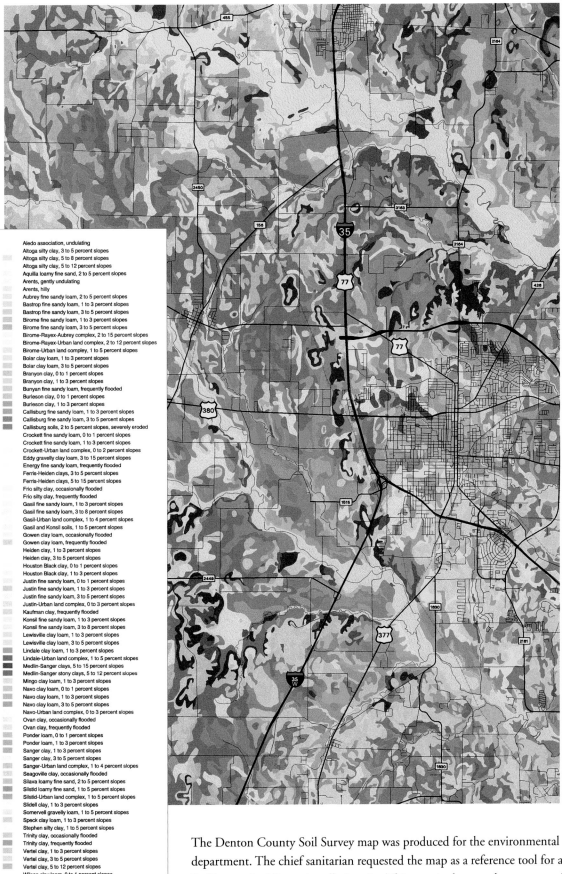

Denton County Planning Department
Denton, Texas

By Rachel Crowe, Mark Lindsey,
John R. Long, Michelle Markham, and
Tim Yarbrough

Contact
John Long
john.long@co.denton.tx.us

Software
ArcInfo 7.2.1
Hardware
Windows NT workstations
Printer
HP DesignJet 750C+
Data Source(s)
Denton County and U.S. Department
of Agriculture-Natural Resource
Conservation Service

Aledo association, undulating
Altoga silty clay, 3 to 5 percent slopes
Altoga silty clay, 5 to 8 percent slopes
Altoga silty clay, 5 to 12 percent slopes
Aquilla loamy fine sand, 2 to 5 percent slopes
Arents, gently undulating
Arents, hilly
Aubrey fine sandy loam, 2 to 5 percent slopes
Bastrop fine sandy loam, 1 to 3 percent slopes
Bastrop fine sandy loam, 3 to 5 percent slopes
Birome fine sandy loam, 1 to 3 percent slopes
Birome fine sandy loam, 3 to 5 percent slopes
Birome-Rayex-Aubrey complex, 2 to 15 percent slopes
Birome-Rayex-Urban land complex, 2 to 12 percent slopes
Birome-Urban land comply, 1 to 5 percent slopes
Bolar clay loam, 1 to 3 percent slopes
Bolar clay loam, 3 to 5 percent slopes
Branyon clay, 0 to 1 percent slopes
Branyon clay, 1 to 3 percent slopes
Bunyan fine sandy loam, frequently flooded
Burleson clay, 0 to 1 percent slopes
Burleson clay, 1 to 3 percent slopes
Callisburg fine sandy loam, 1 to 3 percent slopes
Callisburg fine sandy loam, 3 to 5 percent slopes
Callisburg soils, 2 to 5 percent slopes, severely eroded
Crockett fine sandy loam, 0 to 1 percent slopes
Crockett fine sandy loam, 1 to 3 percent slopes
Crockett-Urban land complex, 0 to 2 percent slopes
Eddy gravelly clay loam, 3 to 15 percent slopes
Energy fine sandy loam, frequently flooded
Ferris-Heiden clays, 3 to 5 percent slopes
Ferris-Heiden clays, 5 to 15 percent slopes
Frio silty clay, occasionally flooded
Frio silty clay, frequently flooded
Gasil fine sandy loam, 1 to 3 percent slopes
Gasil fine sandy loam, 3 to 8 percent slopes
Gasil-Urban land complex, 1 to 4 percent slopes
Gasil and Konsil soils, 1 to 5 percent slopes
Gowen clay loam, occasionally flooded
Gowen clay loam, frequently flooded
Heiden clay, 1 to 3 percent slopes
Heiden clay, 3 to 5 percent slopes
Houston Black clay, 0 to 1 percent slopes
Houston Black clay, 1 to 3 percent slopes
Justin fine sandy loam, 0 to 1 percent slopes
Justin fine sandy loam, 1 to 3 percent slopes
Justin fine sandy loam, 3 to 5 percent slopes
Justin-Urban land complex, 0 to 3 percent slopes
Kaufman clay, frequently flooded
Konsil fine sandy loam, 1 to 3 percent slopes
Konsil fine sandy loam, 3 to 8 percent slopes
Lewisville clay loam, 1 to 3 percent slopes
Lewisville clay loam, 3 to 5 percent slopes
Lindale clay loam, 1 to 3 percent slopes
Lindale-Urban land complex, 1 to 5 percent slopes
Medlin-Sanger clays, 5 to 15 percent slopes
Medlin-Sanger stony clays, 5 to 12 percent slopes
Mingo clay loam, 1 to 3 percent slopes
Navo clay loam, 0 to 1 percent slopes
Navo clay loam, 1 to 3 percent slopes
Navo clay loam, 3 to 5 percent slopes
Navo-Urban land complex, 0 to 3 percent slopes
Ovan clay, occasionally flooded
Ovan clay, frequently flooded
Ponder loam, 0 to 1 percent slopes
Ponder loam, 1 to 3 percent slopes
Sanger clay, 1 to 3 percent slopes
Sanger clay, 3 to 5 percent slopes
Sanger-Urban land complex, 1 to 4 percent slopes
Seagoville clay, occasionally flooded
Silava loamy fine sand, 2 to 5 percent slopes
Silstid loamy fine sand, 1 to 5 percent slopes
Silstid-Urban land complex, 1 to 5 percent slopes
Slidell clay, 1 to 3 percent slopes
Somervell gravelly loam, 1 to 5 percent slopes
Speck clay loam, 1 to 3 percent slopes
Stephen silty clay, 1 to 5 percent slopes
Trinity clay, occasionally flooded
Trinity clay, frequently flooded
Vertel clay, 1 to 3 percent slopes
Vertel clay, 3 to 5 percent slopes
Vertel clay, 5 to 12 percent slopes
Wilson clay loam, 0 to 1 percent slopes
Wilson clay loam, 1 to 3 percent slopes
Wilson-Urban land complex, 0 to 2 percent slopes
Water
Dam

The Denton County Soil Survey map was produced for the environmental division of the county health department. The chief sanitarian requested the map as a reference tool for assessing the type of on-site sewage facility required for an installation site. This map is also popular among real estate companies, developers, consulting firms, and other public and private agencies.

1:1,000-Scale Digital Topographic Map

Survey and Mapping Office
 Lands Department
North Point, Hong Kong

By Survey and Mapping Office, Lands Department

Contact
K.F. Yeung

Software
ArcInfo
Hardware
UNIX workstation
Printer
Electrostatic plotter
Data Source(s)
Survey and Mapping Office, Lands Department, Hong Kong

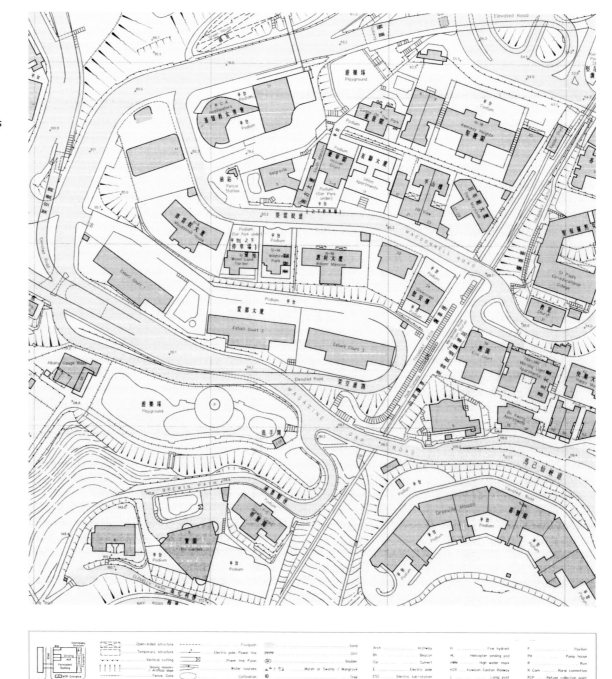

This map illustrates the Central District of Hong Kong at the scale of 1:1,000, which is part of the large-scale base map of Hong Kong. The map is printed from a digital file and contains different layers of map features such as roads, buildings, and utility points.

The Survey and Mapping Office (SMO) of the Lands Department maintains the accuracy of the map data and keeps the map up-to-date. Apart from the large-scale map, which covers the whole territory of Hong Kong, SMO also maintains other maps in scales ranging from 1:5,000 to 1:200,000. Further details are available at www.info.gov.hk/landsd.

Image Chart of Annapolis, Maryland

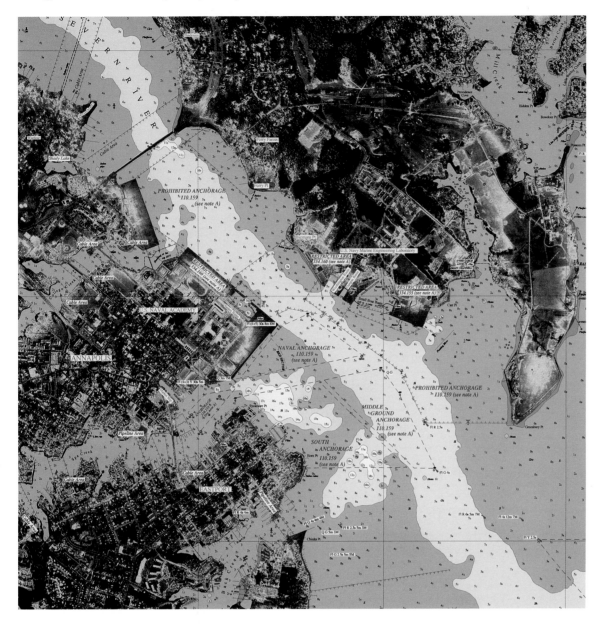

National Imagery and Mapping
 Agency
Bethesda, Maryland

By Stephen M. Lewis and Tom Roff

Contact
Stephen Lewis
lewiss2@nima.mil

Software
ArcInfo 7.2.1 and ERDAS IMAGINE
Hardware
Sun SPARC 20 workstation
Printer
HP DesignJet 755CM
Data Source(s)
Digital Nautical Chart and U.S.
Geological Survey orthophoto

73 | military

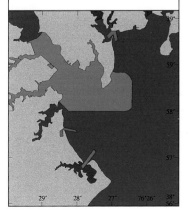

The Image Chart of Annapolis, Maryland, is a prototype product produced by the Hydrographic National Imagery and Mapping Agency (NIMA) Production Cell (Hydro NPC). The chart combines NIMA's Digital Nautical Chart (DNC) vector data with imagery to produce a new version of the standard NIMA nautical chart.

In the new image chart, the land areas and cultural features are replaced by one-meter imagery (U.S. Geological Survey digital orthophoto quadrangle). All other features and symbology are identical to the standard nautical chart. In addition, information from the National Ocean Service publication *Coast Pilot* is included in the margin.

Determination of Obstacles in the Vicinity of Aerodromes Using Obstacle Limitation Surfaces and GIS

General Command of Mapping
 Cartography Department
Ankara, Turkey

By Ali Ulubay

Contact
Ali Ulubay
aulubay@hgk.mil.tr

Software
ArcInfo 7.1.2, ARC GRID, ArcView GIS
3.1, ArcView 3D Analyst 1, ArcView
Spatial Analyst 1, and Freehand
Version 8.0
Hardware
Alpha 255/300, Pentium Pro 200, and
Power Mac G3-600
Printer
OCE 5350
Data Source(s)
International Civil Aviation
Organization parameters for obstacle
limitation surfaces and
1:25,000-scale elevation,
hydrography, geology, industry,
transportation, vegetation,
population, and utilities maps

DIGITAL TERRAIN MODEL

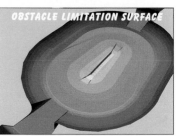

OBSTACLE LIMITATION SURFACE

BUILDINGS AND CONSTRUCTIONS

GEOLOGY

VEGETATION

It is important to maintain obstacle-free airspace around airports; man-made objects such as buildings, antennas, and towers, as well as landscape features, can pose dangers to aircraft leaving or approaching airfields. This project helped establish a system for identifying potentially hazardous obstacles near an airport and querying their attributes. It will be used for analysis in determining airfield locations and monitoring and assisting in regional planning around airports.

The International Civil Aviation Organization (ICAO) has established guidelines called obstacle limitation surfaces (OLSs) for maintaining obstacle-free airspace in the vicinity of airports. These OLSs, combined with geology, terrain, vegetation, and land use data sets, were converted to points on a digital terrain model (DTM) using ARC Macro Language (AML) in an ArcInfo environment.

Contours of six digital topographic maps were used for the DTM of the area, and triangulated irregular networks of the terrain and OLSs were converted to LATTICE for surface comparisons and for other analysis in the ARC GRID environment. Features were digitized from 1:25,000-scale sheets, and attributes were collected for the GIS. ARC GRID analyzed slope, visibility, and hydrology, and the final maps were prepared in ArcView 3D Analyst.

Comparison of Stream Extent and Location from Several Data Sources

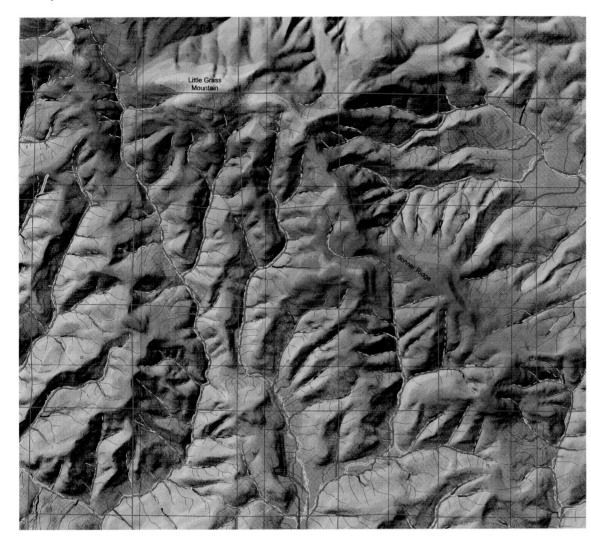

Oregon Department of Forestry
Salem, Oregon

By Robert Nall

Contact
Robert Nall
rnall@odf.state.or.us

Software
ArcView GIS 3.1 and ArcView Spatial
Analyst
Hardware
PC
Printer
HP DesignJet 450C
Data Source(s)
U.S. Geological Survey 10-meter
digital elevation models

Quadrangle Streams
Orthophoto Streams
Fish Bearing Streams
Streams Based on Drainage Area
100 Acre Basins
60 Acre Basins
50 Acre Basins
40 Acre Basins
30 Acre Basins
20 Acre Basins
10 Acre Basins

Northwest Oregon Counties

Whether a stream contains fish and if it is perennial or seasonal are primary factors in determining forest management strategies. The State Forest Management Program of the Oregon Department of Forestry needs this information to classify land according to the type of management that will take place there. The Northwest Oregon State Forest Management Plan and the Western Oregon Habitat Conservation Plan, currently under development, will include strategies for managing aquatic and riparian areas based on those attributes.

Because data for fish presence was incomplete and other data on streams varied, the department began this mapping project to compare three different data sources for hydrologic information and select the most appropriate one for use in the development of the two plans.

The best available data sources included a hydrology theme digitized from the U.S. Geological Survey (USGS) Summit quadrangle at 1:24,000 scale, a hydrology theme digitized from 1994 color orthophotos at 1:12,000-scale, and hydrology themes derived from 10-meter digital elevation models. Field checks of the data showed that the quadrangles do not include many existing streams, while the orthophoto data indicated far too many streams.

ArcView Spatial Analyst and hydrology extensions helped analyze the digital elevation model to determine the location and extent of streams. Watershed (basin) size in acres (10, 20, 30, 40, 50, 60, and 100) was the basis for determining the extent of the streams.

Regional Water Planning Groups and Water Resources of Texas

Texas Water Development Board
Austin, Texas

By Mark Hayes and Teresa Howard

Contact
Mark Hayes
mark.hayes@twdb.state.tx.us

Software
ArcInfo 7.2.1
Hardware
Sun SPARC 5 workstation
Printer
HP DesignJet 650C
Data Source(s)
Texas Water Development Board GIS
database, U.S. Geological Survey
digital line graphs and hydrologic
units, and U.S. Census Bureau TIGER
county boundaries

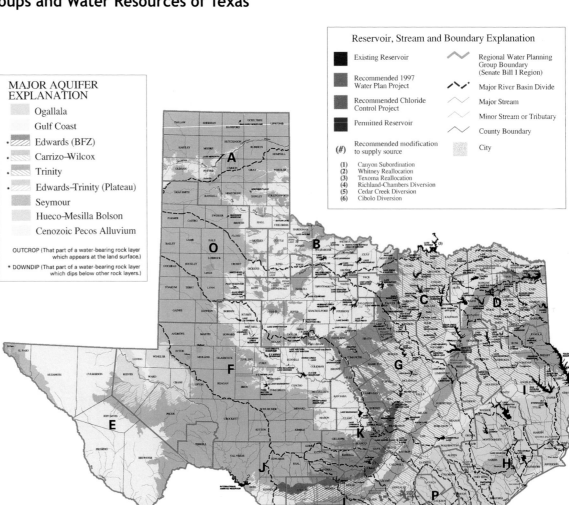

MAJOR AQUIFER EXPLANATION

Ogallala
Gulf Coast
• Edwards (BFZ)
• Carrizo–Wilcox
• Trinity
• Edwards–Trinity (Plateau)
Seymour
Hueco–Mesilla Bolson
Cenozoic Pecos Alluvium

OUTCROP (That part of a water-bearing rock layer which appears at the land surface.)

* DOWNDIP (That part of a water-bearing rock layer which dips below other rock layers.)

Reservoir, Stream and Boundary Explanation

Existing Reservoir
Recommended 1997 Water Plan Project
Recommended Chloride Control Project
Permitted Reservoir

(#) Recommended modification to supply source

(1) Canyon Subordination
(2) Whitney Reallocation
(3) Texoma Reallocation
(4) Richland-Chambers Diversion
(5) Cedar Creek Diversion
(6) Cibolo Diversion

Regional Water Planning Group Boundary (Senate Bill I Region)
Major River Basin Divide
Major Stream
Minor Stream or Tributary
County Boundary
City

In June 1997, Senate Bill 1, comprehensive water legislation enacted by the 75th Texas legislature, was signed into law. The legislation put into place a regional water planning process designed to meet the water needs of Texas as the state faces both increasing population and ongoing vulnerability to drought. In 1998, following the guidelines established by the bill, the Texas Water Development Board adopted state and regional water planning rules, delineated 16 regional planning areas, and selected members to serve in each regional water planning group.

GIS analysis played an important role in the regional planning area delineation. Senate Bill 1 called for the delineation of regional planning areas based on factors such as river basin and aquifer characteristics, water utility development patterns, socioeconomic characteristics, and political subdivision boundaries. These planning areas grew out of the combined inputs from experienced water planning professionals, GIS analysis, and public comment.

The map, Regional Water Planning Groups and Water Resources of Texas, illustrates the context of each planning area. Depicted are political boundaries, city locations, major river basin divides, as well as surface and groundwater resources. Existing and proposed reservoirs are displayed thematically. The challenge to the cartographers who designed this map was to present many layers of information in a legible and visually pleasing way to facilitate the regional water planning process for both water professionals and laypeople.

1999 Deer Zone Map

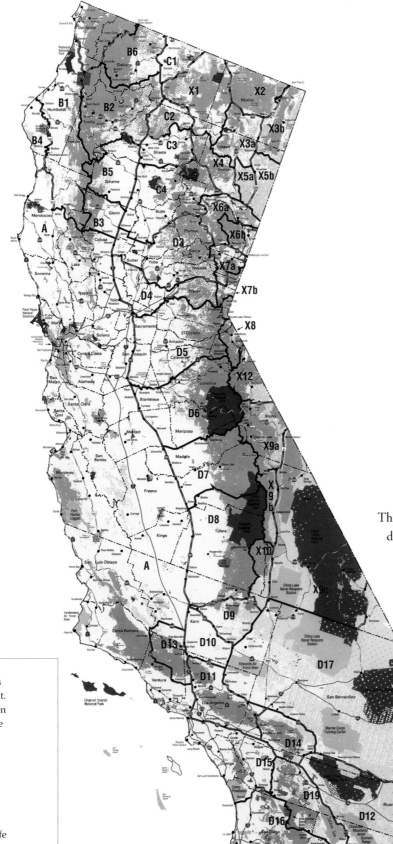

California Department of Fish and
 Game
Sacramento, California

By Thomas Lupo

Contact
Thomas Lupo
tom_lupo@dot.ca.gov

Software
ArcInfo 7.2.1 and ARCPLOT
Hardware
Windows NT workstations
Printer
Printed commercially on a four-color
offset printing press and folded to
order
Data Source(s)
California Department of Fish and
Game, Teale Data Center

This map is one of 5,000 copies printed for
distribution to hunting license agents and large
sporting goods retail outlets throughout
California. The electronic prepress files prepared
for the printing firm were in an Adobe
Illustrator format file created from the
map composition using the Arc Illustrator
command. The resolution of the
resulting printed map is far superior to
what could have been created using
a standard 300 dpi ink-jet plotter.
In this process, the text labels can
be much smaller than what would
normally be used on a plotted
map.

Legend

- Bur. Indian Affairs
- Bur. of Land Mgmt.
- Bur. of Reclamation
- Calif. Fish & Game
- Calif. Forestry
- Calif. Parks & Rec
- Military
- Natl. Park Service
- Other gov't lands
- Private
- Private Preserve
- U.S. Fish & Wildlife
- U.S. Forest Service
- Water Body

N Zone Boundaries and Codes

Incorporating the Protection of Biodiversity into County Land Use Planning—A Gap Analysis Pilot Project in Pierce County, Washington

University of Washington
Seattle, Washington

*By Doug Pflugh, Wood Turner, and
Washington Gap Analysis Project
staff*

Contact
Dr. Frank Westerlund
fwest@u.washington.edu

Software
ArcView GIS 3.1 with data generated
in ArcInfo 7.1.2
Hardware
Gateway G6-200 and Sun Ultra 2
workstation
Printer
HP DesignJet 650C
Data Source(s)
Pierce County GIS; Washington State
Gap Analysis Project; Washington
Department of Natural Resources;
Washington Department of Fish and
Wildlife; and Remote Sensing
Applications Laboratory/Department
of Urban Design and Planning,
University of Washington

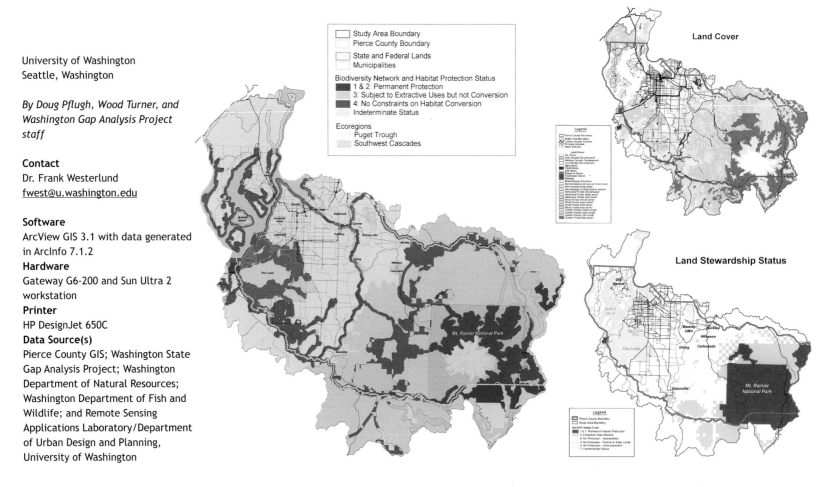

Gap analysis is a process by which areas of high conservation priority are identified. The process uses land cover, wildlife/habitat relationship models, and other data to predict the distribution of wildlife species in a given geographic area. Gap analysis was designed to be a proactive approach to conservation that identifies important habitat areas or species before they become threatened by habitat degradation or loss. Conservation of critical habitat areas supports the ecological integrity of an area by protecting its natural biodiversity, the range of living organisms and the processes that create and sustain them.

The Washington Gap Analysis Project (WAGAP) has completed a gap analysis of Washington, but additional work is needed to make this process an effective tool for local governments. Although the data from WAGAP has been summarized and analyzed at a statewide level, this analysis was conducted using large-scale ecoregions and vegetation zones, which are not easily applied to county land use planning and habitat conservation.

The Department of Urban Design and Planning (UDP) at the University of Washington (through its remote sensing applications laboratory), in cooperation with WAGAP (at the Washington Cooperative Fish and Wildlife Research Unit), is conducting an ongoing program to develop procedures and materials supporting the application of WAGAP data and methodology to local land use planning.

These maps document part of the work performed during Phase I of the Pierce County pilot. The project's objectives were to map land cover; model the distributions of terrestrial vertebrates; identify land cover types, vertebrate species, and areas of high vertebrate species richness inadequately represented in protected areas; and make this data available to users in a readily accessible format.

The Land Stewardship Status map depicts the conservation status classification for the study area.

Current land cover within each vegetation zone was mapped using 1991 Landsat satellite Thematic Mapper imagery at a nominal 100-hectare minimum-mapping unit. The Land Cover map shows the land cover classification for the study area. This classification was the foundation of the predicted species distribution modeling performed for the pilot project.

The final biodiversity network and its breakdown by land stewardship status are shown on the Biodiversity Network and Habitat Protection Status map.

Minnesota Land Use and Cover: A 1990 Census of the Land

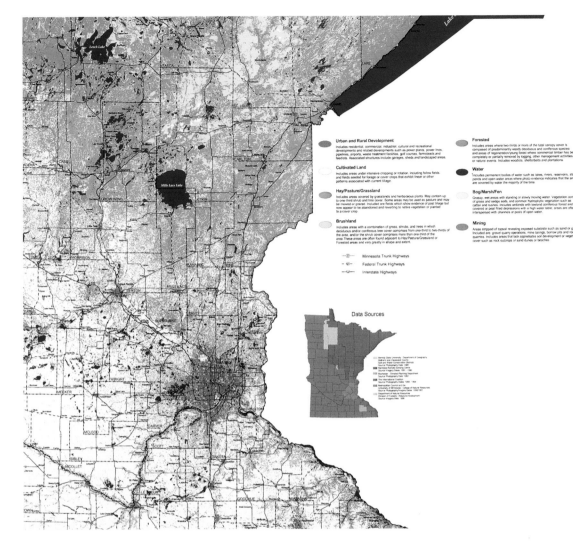

Minnesota Department of Natural Resources
St. Paul, Minnesota

By Timothy N. Loesch

Contact
Timothy Loesch
tim.loesch@dnr.state.mn.us

Software
ArcView GIS and ArcView Spatial Analyst
Hardware
DELL Precision 410 Windows NT workstation
Printer
Professional four-color process
Data Source(s)
Bemidji State University, Manitoba Remote Sensing Centre, Rochester-Olmsted Planning Department, International Coalition Source Photography, Metropolitan Council, University of Minnesota, and Minnesota Department of Natural Resources Division of Forestry

Developed with ArcView GIS 3.1 and ArcView Spatial Analyst software, this map shows land cover and land use in the state of Minnesota. Its title, "Minnesota Land Use and Cover: A 1990 Census of the Land," was chosen to reflect the variety of data sources that contributed to its production. A series of land use and cover data collection projects was conducted from 1990 to 1996 and resulted in the statewide coverage seen on this map. The data presented a considerable challenge in reconciling the differences in classification systems.

Three of North America's ecological regions, or biomes, representing major climate zones converge in Minnesota—prairie parkland, deciduous forest, and coniferous forest. The presence of three biomes in one non-mountainous state is unusual and accounts for the diversity of ecological communities in Minnesota. These three zones are clearly visible on this map.

The Prairie Parkland province covers most of the western and southern sections of Minnesota and was occupied by tallgrass prairie before settlement. The topography is predominantly level to gently rolling, and major landforms include lake plains and ground moraines.

The Laurentian Mixed Forest province comprises the true forested region of Minnesota. It consisted of continuous conifer, conifer–hardwood mix, or hardwood forest vegetation before settlement with variable topography. Landforms range from lake plains and outwash plains to ground and end moraines.

The Eastern Broadleaf Forest province bridges the transition zone between prairie to the west and true forest to the east. Topography varies from level in the plains to very steep in the Paleozoic plateau subsection of the southeast. Major landforms include lake plains; outwash plains; end moraines; ground moraines; drumlin fields; and unglaciated, dissected sedimentary beds.

The Legislative Commission on Minnesota Resources (LCMR) provided principal funding for this project. Map sponsors included the Association of Minnesota Counties, the University of Minnesota Center for Urban and Regional Affairs (CURA), the Science Museum of Minnesota, and the Minnesota Department of Natural Resources.

Fragmented Forests in the Chesapeake Bay Watershed: Large Contiguous Forest Patches

Chesapeake Bay Program/University
  of Maryland Eastern Shore
Annapolis, Maryland

By Howard Weinberg

Contact
Howard Weinberg
hweinber@chesapeakebay.net

Software
ArcInfo 7.2 and ArcView GIS 3.1
Hardware
Dell PC
Printer
HP DesignJet 750C+
Data Source(s)
Multi-Resolution Land Characteristics
land cover data set, U.S. Geological
Survey, and State Breeding Bird
atlases

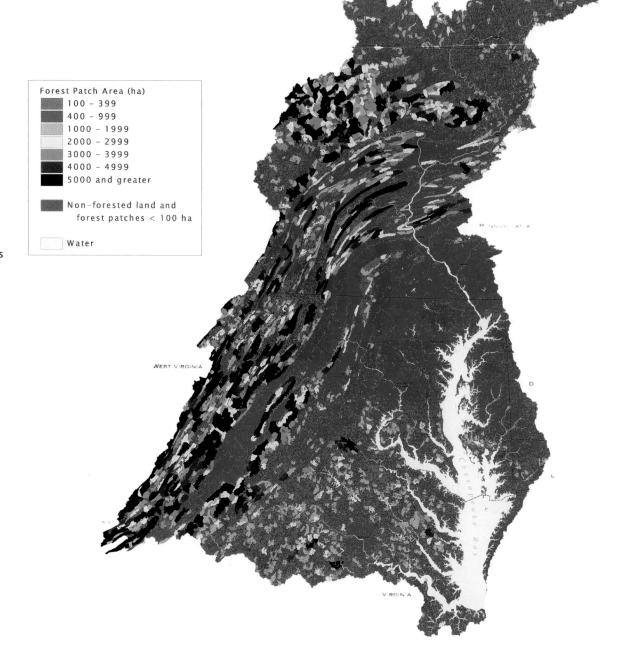

Forest Patch Area (ha)
100 – 399
400 – 999
1000 – 1999
2000 – 2999
3000 – 3999
4000 – 4999
5000 and greater

Non-forested land and
forest patches < 100 ha

Water

Two complementary maps were created for use in evaluating habitat for forest interior dwelling neotropical migrant songbirds.

Forest fragmentation is increasing in the Chesapeake Bay watershed due to increasing development in suburban and rural areas and the clearing of land for agriculture. The Large Contiguous Forest Patches map shows a mosaic of patches of varying sizes. The largest patches are concentrated in rural areas with higher elevation and steeper slopes. Research has suggested that 3,000 hectares or more of contiguous forest are necessary for the maximum probability of occurrence of many neotropical species.

Another map that was produced, the Habitat of Forest Interior Dwelling Birds map, shows considerably fewer patches of large interior forest available to those birds requiring contiguous forest for successful breeding.

Landscape-Level Forest Planning at the McGregor Model Forest Association

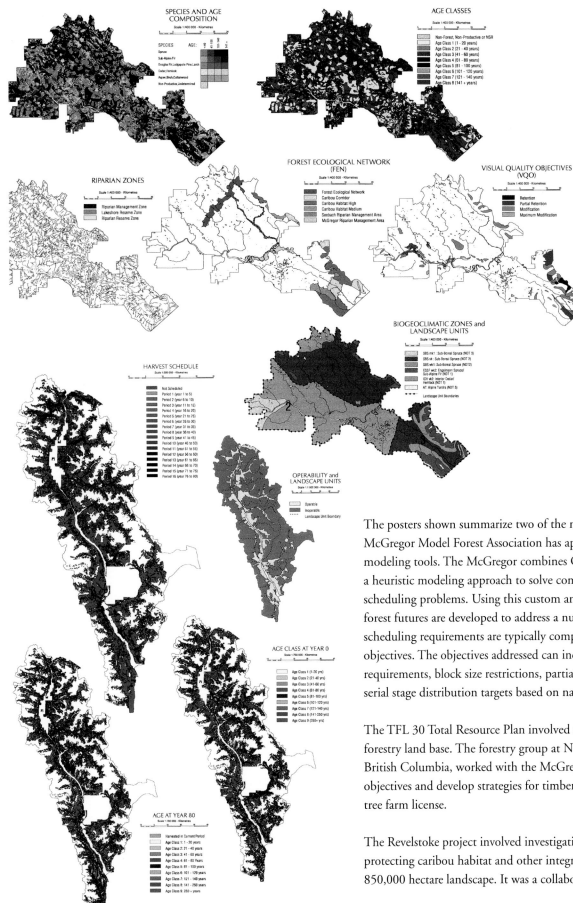

McGregor Model Forest Association
Prince George, British Columbia

*By Carey Lockwood, Bruce McLellan,
Don Morgan, Tom Moore, and
Kate Sherren*

Contact
Kate Sherren
kate@mcgregor.bc.ca

Software
ArcInfo 7.2.1, ARCEDIT, ARCPLOT,
ARC TIN, ARC GRID, Custom Model,
LaTeX, and Adobe Illustrator
Hardware
Sun Ultra 2 workstations
Printer
HP DesignJet 755CM and 750C+
Data Source(s)
British Columbia Ministry of
Environment, Lands and Parks Forest
Inventory, TRIM Digital Elevation
Model, growth and yield curves
generated from standard British
Columbia Ministry of Forests VDYP
and TIPSY models, and licensee-
supplied harvesting constraint layers
and custom model software outputs

The posters shown summarize two of the many case studies to which the
McGregor Model Forest Association has applied their spatial timber supply
modeling tools. The McGregor combines GIS, growth and yield data, and
a heuristic modeling approach to solve complex spatial and temporal forest
scheduling problems. Using this custom analytical planning model, alternative
forest futures are developed to address a number of management scenarios. The
scheduling requirements are typically composed of large numbers of interacting
objectives. The objectives addressed can include block adjacency green-up
requirements, block size restrictions, partial cutting targets for riparian areas, and
serial stage distribution targets based on natural disturbance patterns.

The TFL 30 Total Resource Plan involved strategic planning on an operational
forestry land base. The forestry group at Northwood, Inc., in Prince George,
British Columbia, worked with the McGregor Model Forest Association to define
objectives and develop strategies for timber supply analysis of this 181,000 hectare
tree farm license.

The Revelstoke project involved investigating the impacts to timber supply of
protecting caribou habitat and other integrated resource management areas on an
850,000 hectare landscape. It was a collaborative forest districtwide study.

Lake Tahoe Region Wildland Fire Susceptibility

Jones & Stokes
Sacramento, California

By Kelly Berger, Bruce Boyd, and Chris DiDio

Contact
Chris DiDio
chris@jsanet.com

Software
ArcInfo, ARC GRID, ArcView
3D Analyst, FARSIT, and FLAMMAP
Hardware
Windows NT workstation
Printer
HP DesignJet 2500CP
Data Source(s)
USDA Forest Service, Lake Tahoe Basin Management Unit, El Dorado National Forest, and Tahoe National Forest

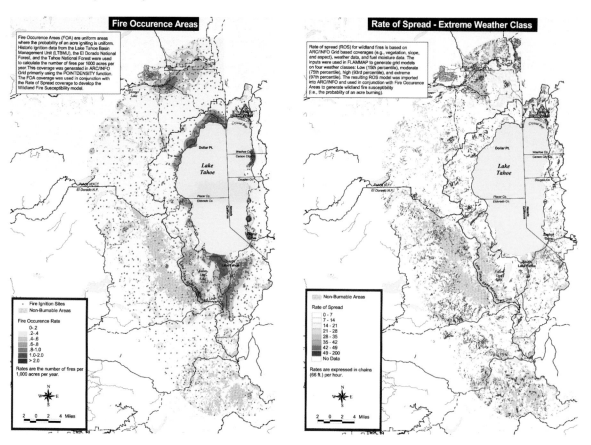

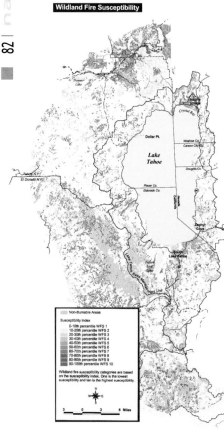

Wildland fire susceptibility is dependent on many factors including weather, topography, and fuels. The probability of any given acre burning in the Lake Tahoe basin was modeled using GIS data, ArcInfo, and the fire behavior modeling programs FARSIT and FLAMMAP. This map illustrates historical fire occurrence areas, rate of spread for a potential fire, and a composite wildland fire susceptibility index. The index is spatially represented on the wildland fire susceptibility map.

Fire occurrence areas (FOAs) are uniform areas where the probability of an acre igniting is uniform. With each unique combination of an FOA (probability of an acre igniting) and rate of spread (fire behavior), a wildland fire susceptibility index was calculated for each 30-meter cell in the grid. Areas that are covered by water or are barren were eliminated from the total burnable acres within each FOA. The susceptibility index is calculated by dividing the expected acres burned (final fire size) by the total acres in the FOA.

Historic ignition data was used to calculate the number of fires per 1,000 acres per year. The FOA coverage was used in conjunction with the rate of spread coverage to develop the wildfire susceptibility model.

Rate of spread (ROS) for wildland fires is based on coverages (e.g., vegetation, slope, and aspect), weather data, and fuel moisture data. The ROS model was imported into ArcInfo and used in conjunction with FOA to generate wildland fire susceptibility (i.e., the probability of an acre burning).

Hidden River Cave and the Longest Cave

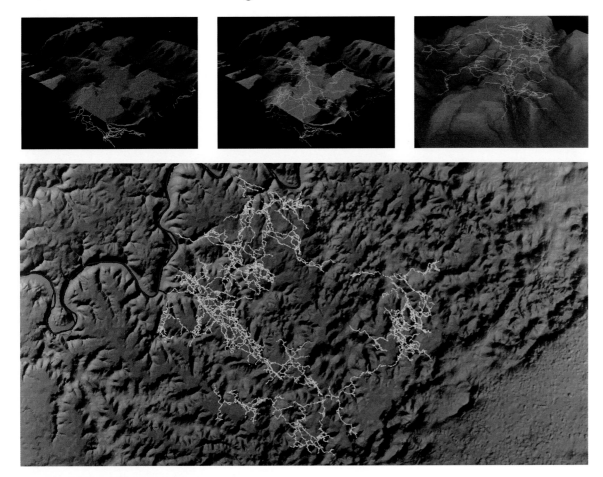

Cave Research Foundation
Frankfort, Kentucky

By Mike Yocum

Contact
Mike Yocum
myocum@mis.net

Software
ArcView GIS 3.1, ArcView Spatial
Analyst 1.1, and ArcView 3D Analyst 1
Hardware
Micron Pentium Pro PC
Printer
HP DesignJet 750C+
Data Source(s)
Cave Research Foundation, Central
Kentucky Karst Coalition, and U.S.
Geological Survey

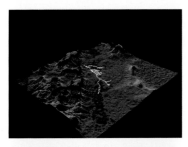

The Cave Research Foundation (CRF) is dedicated to facilitating research, management, and the interpretation of caves and karst; forming partnerships to protect, preserve, understand, and interpret caves and karst; and promoting the long-term conservation of caves and karst ecosystems.

Understanding, protecting, managing, and conserving such an extraordinarily rich and complex set of resources require tools capable of integrating, manipulating, and querying the information used to describe their many facets. In 1997, the CRF established a GIS Resource Development Program. The goal of the program is to assist CRF personnel, federal agency staff, and other researchers to access and use spatial data, GIS applications, and other software tools for the purpose of cave and karst resource management. A longer-term goal is to use GIS to develop a collective knowledge and support base for cave conservation, protection, and management.

The Hidden River Cave maps to the left were created by CRF's GIS Resource Development Program to celebrate GIS Day 1999. They show (1) a three-dimensional view of Hidden River Cave in its topological setting, (2) sources of waste that polluted the cave, and (3) the location of cave survey stations in the southern part of the system. Hidden River Cave underlies the small Kentucky town of Horse Cave.

The three-dimensional maps at the top of the page illustrate the complexity of passages in Mammoth Cave and their relationship to surface terrain. The large map shows the entire Mammoth Cave system located at the edge of the Dripping Springs escarpment in south central Kentucky. With almost 350 miles surveyed, Mammoth Cave is the longest cave in the world. Survey data for the cave passages has been collected for more than 40 years by CRF working in cooperation with the National Park Service at Mammoth Cave National Park.

Submarine Topography Offshore Central California

Monterey Bay Aquarium Research
 Institute (MBARI)
Moss Landing, California

By Norman Maher

Contact
Norman Maher
nmaher@mbari.org

Software
ArcInfo 7, ArcView GIS 3.5, Adobe
Illustrator Version 8.0, and Photoshop
Version 5.0
Hardware
HP Kayak 450 XU, SGI O2, and
Macintosh G3
Printer
HP DesignJet 3500CP
Data Source(s)
National Oceanic and Atmospheric
Administration, U.S. Geological
Survey, and MBARI

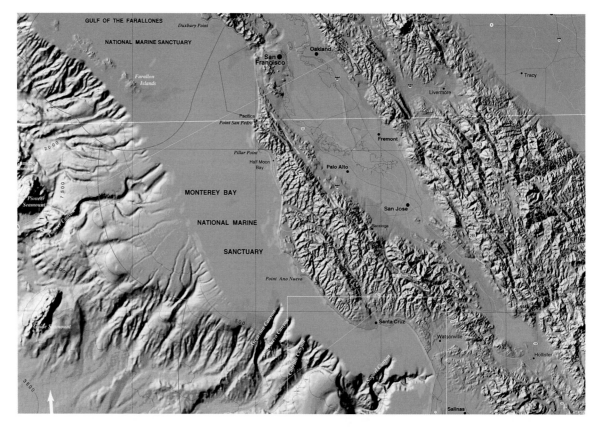

The Monterey Bay National Marine Sanctuary encompasses an offshore region that is marked by a relatively flat and gently sloping continental shelf and a steep and rugged continental slope that is deeply cut by submarine canyons.

The submarine canyons form by erosion similar to the way river canyons on land are formed, except that the erosive forces that carve them are in the form of seafloor turbidity currents (dense sediment-laden bottom currents that periodically occur following catastrophic slope failures). Earthquakes, severe storms, or tidal surges can trigger these failures. Sediments deposited downstream of the canyon mouths form large submarine fan systems on the continental rise and out onto the abyssal plain.

Three volcanic seamounts, Pioneer, Guide, and Davidson, rise above the seafloor west of the sanctuary boundary. The largest of these, Pioneer Seamount, is nearly comparable in size to Mount Shasta. Guide Seamount rises 4,200 feet above the surrounding seafloor, and Pioneer Seamount rises 6,300 feet above the surrounding seafloor. These submarine volcanoes were formed 12 to 14 million years ago.

The high-resolution shaded-relief image of Monterey Canyon reveals details such as large submarine landslides and numerous gullies cutting the steep sides of the canyon. The distinct shape of the canyon is due to fault displacement that altered the canyon's course, preferential erosion along weak fault zone rocks, and large slumps and slides. Red dashed lines indicate approximate traces of the San Gregorio and Monterey Bay fault zones. Carmel Canyon is relatively straight in comparison to Monterey Canyon, as its axis follows the weaker, more easily eroded rocks along a linear fault zone.

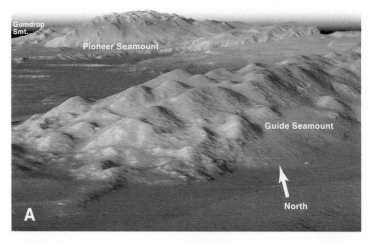

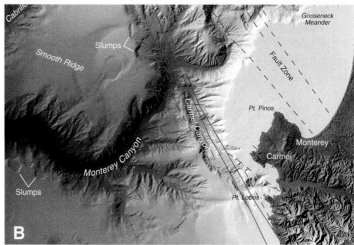

GIS Applications to Maritime Boundary Definitions

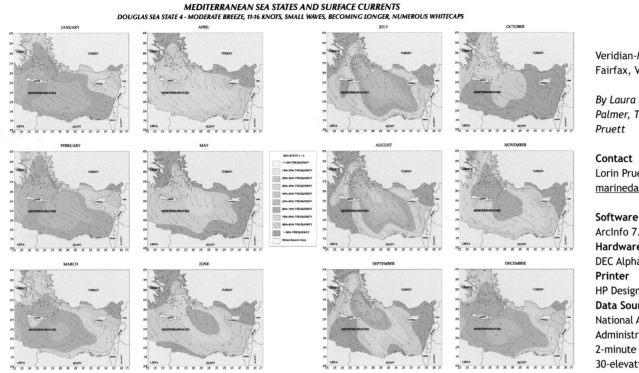

MEDITERRANEAN SEA STATES AND SURFACE CURRENTS
DOUGLAS SEA STATE 4 - MODERATE BREEZE, 11-16 KNOTS, SMALL WAVES, BECOMING LONGER, NUMEROUS WHITECAPS

Veridian-MRJ Technology Solutions
Fairfax, Virginia

By Laura (Lu) Crenshaw, Dr. Hal Palmer, Tracie Penman, and Lorin Pruett

Contact
Lorin Pruett
marinedata@mrj.com

Software
ArcInfo 7.2.1
Hardware
DEC Alpha/UNIX platform
Printer
HP DesignJet 650C
Data Source(s)
National Aeronautics and Space Administration bathymetry, Topex 2-minute bathymetry, GTOPO 30-elevation, ArcWorld 3-World Vector Shoreline, and U.S. Navy Marine Climatic Atlas of the World, 1976

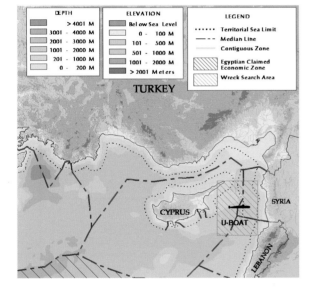

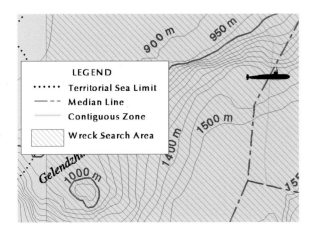

Recent advances in deepwater search and salvage have enabled access to shipwrecks that were once thought to be inaccessible. The well-publicized search and location of the Titanic is one example. The primary factor in successfully locating a wreck is searching in the geographically correct area. A "search box" is defined through the synergism of multiple and sometimes apparently contradictory information.

For example, during World War II British destroyers forced the German submarine U-559 to the surface and a boarding party recovered vital code books and equipment before the submarine sank. While the position of the wreck was presumed from the destroyer's location, subsequent analyses of surface currents, sea state, and prevailing winds at that time indicated a more northerly search box was probable. Other strategic devices went down with the submarine, but the wreck site was not investigated.

Use of a GIS to plan wreck searches helps to efficiently organize and present the varying data sources and ultimately to define the search box. When it is defined, the characteristics of the bottom, including depths, bottom sediments, and other area features such as underwater canyons or known debris, help the side scan sonar operators plan the two-direction, towfish altitude and potential for false targets. Time spent at sea, specifically in a deepwater search, is expensive (sometimes exceeding $40,000/day), and a tool that enables review of all data spatially can contribute to a successful and cost-effective search.

Ownership and the right to recover antiquities and objects of value from the seafloor remain controversial issues. Knowledge of the extent of maritime boundaries and a coastal state's jurisdiction over waters and the seabed are essential to ensure that those who seek to salvage such objects conform to legal protocols. Veridian-MRJ has produced and is maintaining the Global Maritime Boundary Database CD-ROM, which provides much of the information requisite to the analyses of rights and limits for salvors and marine archaeologists.

Utilizing ARCOView Extension for Exploration Mapping

ARCO International Oil and Gas
 Company
Plano, Texas

By Scott Sitzman

Contact
Scott Sitzman
ssitzma@mail.arco.com

Software
ArcView GIS 3.0b and ARCOView
extension
Hardware
Sun UNIX workstation
Printer
HP DesignJet 2000CP
Data Source(s)
Earth Science Associates, Geologic
Data Systems, IHS Energy Group, and
Scripps Institute

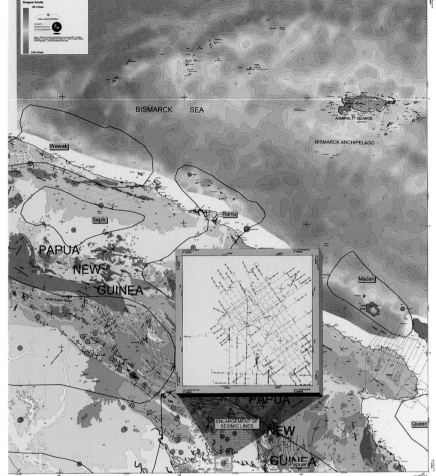

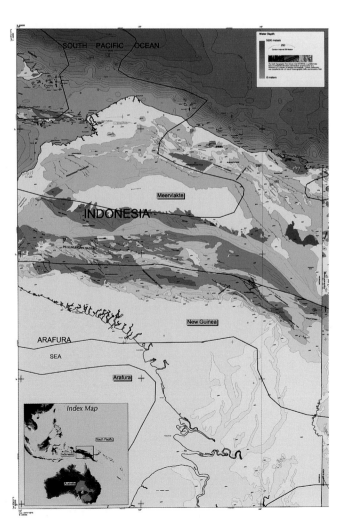

These maps demonstrate ARCOView's mapping capabilities. ARCO internally funded and developed ARCOView with assistance from Eagle Information Mapping of Bellaire, Texas, and Idea Integration of Dallas, Texas. ARCOView is an extension that enables ArcView GIS to serve as the database management and mapping system for the exploration, engineering, and environmental departments of ARCO.

The ARCOView extension provides five areas of enhancements to ArcView GIS. They are necessities in the petroleum industry performing these functions: (1) standardized, simple map generation for the casual user; (2) integration with currently used software; (3) data import/export capabilities; (4) seismic navigational data management; and (5) miscellaneous utilities for projection and data conversion/automation.

Oil and Gas Map Series

Koch Industries—LPG/NGL Operations

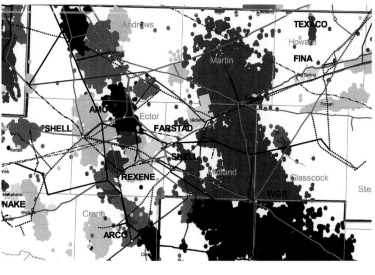

Oil and Gas Production in 1998/Changes in Oil and Gas Production: 1988-1998

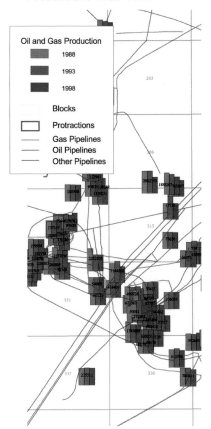

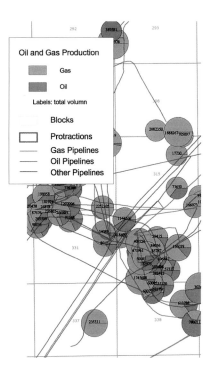

Koch Industries—LPG/NGL Operations

Koch's gas liquids group of companies employs an extensive system of fractionators, propane terminals, underground storage facilities, and thousands of miles of pipeline for gathering and distributing natural gas liquids. One new business in which this group participates is Diamond-Koch, LLC. This LLC offers customers a more efficient source for transportation, storage, processing, and distribution for both natural gas and refinery liquids and petrochemical feedstock. This map was created to study LPG/NGL business opportunities in west Texas. The map displays Koch and competitor LPG pipelines, gas processing plants, refineries, and fractionation facilities, along with oil and gas field production data.

Oil and Gas Production in 1998/
Changes in Oil and Gas Production: 1988–1998

The U.S. Department of the Interior Minerals Management Service (MMS) is the federal agency responsible for managing the mineral resources on the outer continental shelf (OCS) in an environmentally sound and safe manner. Spatial data is essential to MMS in its efforts to successfully accomplish its mission. The MMS technical information management system (TIMS) database consists of two components, spatial and nonspatial. ArcView GIS and ArcInfo are the analytical and mapping tools for the system. These two maps showing the oil and gas productivity in the Eugene Island arena in the Gulf of Mexico are examples of integration of spatial data in the Spatial Database Engine™ (SDE®) database and nonspatial data in the Oracle database. The TIMS keeps the monthly oil and gas volume production data for each oil and gas well on the OCS in its Oracle database and the spatial position in its SDE database. These kinds of maps are very useful for monitoring, evaluating, and regulating the oil and gas production activities on the OCS. They demonstrate that the power of SDE and ArcView GIS in integration of spatial and nonspatial data enables the easy production of comprehensive maps like these two.

(left)
Koch Asset Mapping
  Koch Industries Inc.
Wichita, Kansas

By Matt Hunsberger, Ed Mann,
John Simmonds, Scott Warwick,
Carl Weisbender, Mark Wiggins,
Ed Wilson, and Murry York

Contact
Scott Warwick
warwicks@kochind.com

Software
ArcInfo 7.2.1, ArcPress, and
ArcView GIS 3.1
Hardware
Dell P300 workstation
Printer
HP DesignJet 750C+
Data Source(s)
Pennwell MapSearch competitor
pipeline, PI/D FieldMap, Wessex, and
Koch Asset Mapping

(far left)
United States Department of the
  Interior, Minerals Management
  Service
New Orleans, Louisiana

By M. Aurand, L. Coats, M. Daigle,
K. Huang, Xueqiao Huang, S. Long,
K. Li, P. Rasmus, and J. Wu

Contact
Xueqiao Huang
xueqiao.huang@mms.gov

Software
ArcInfo 7.2.1, ArcView GIS 3.1, and
SDE 3.0.2
Hardware
PC
Printer
HP DesignJet 750C+

Pipeline Map Series

El Paso Energy Corporation
Houston, Texas

By Shu Gutlek, Gary Hoover, Walter Kronenberger, and Buddy Nagel

Contact
Buddy Nagel
nagelb@epenergy.com

Software
ArcInfo 7.2, ArcView GIS 3.1, ARCEDIT, Microsoft Access, Barco Graphics RIP, and SDE 3
Hardware
Compaq Windows NT workstation
Printer
ENCAD Pro50 and ENCAD Pro60e
Data Source(s)
U.S. Department of Interior, Minerals Management Service; MAPSearch; and Business Location Research; MAPSearch and Geographic Data Technology

(bottom right)
PennWell MAPSearch
Durango, Colorado

By Andrew Bicknell, Jason Kord, and Diane Williams

Contact
Diane Williams
dianew@mapsearch.com

Software
ArcInfo 7.2.1, ARCEDIT, Microsoft Access, and Barco Graphics RIP
Hardware
Windows NT PC, four-color Filmwriter
Plotter
Offset-printed
Data Source(s)
PennWell MAPSearch Research Department and base data from 1:2 million scale U.S. Geological Survey digital line graph files

El Paso Energy Facilities

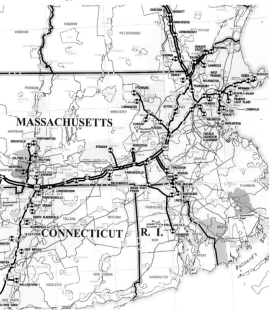

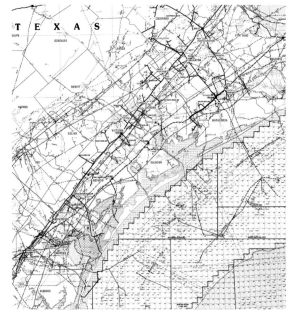

El Paso Energy Facilities

A "map-central" approach to database design assists El Paso Energy in achieving consistent, high-quality products. The El Paso Energy Facilities, Texas Gulf Coast and Offshore Areas, and Northern Division with Local Distribution Companies maps were derived from existing projects and modified (additional scripts, themes, and tables). The maps have a combination of shapefiles and coverages. Text was placed using ArcInfo annotation and by using the label tools within ArcView GIS. Masking of annotation and labels was provided by a set of fonts designed for masking of features. Themes were placed in the correct order for proper display with views added to the active layout where titles, scales, and legends were incorporated to complete the map. PostScript files were generated for imaging and proofing.

LPG/NGL Pipeline and Facilities Map for the United States and Canada

This map contains data gathered by PennWell MAPSearch's Research Department. The pipeline information represented includes ownership, diameter, commodity specification, and direction of product flow. Facility data includes ownership, product specifications, and facility type. The map contains information for more than 61,000 miles of pipeline, 3,500 facilities, and 180 interconnects (where two or more companies exchange commodity) throughout the continental United States and part of Canada.

The base data for the map came from U.S. Geological Survey digital line graph files. State and county polygons were edited and built with line topology to eliminate duplicate arcs along common borders. In highly congested areas, larger scale inset maps provide a more detailed view.

ArcInfo and Microsoft Access maintained the attribute data for the pipelines, facilities, and interconnects. A series of relates pulled the required information from the appropriate database. The data was concatenated into properly formatted text strings and then captured as annotations. ARCEDIT manually edited final annotation placement.

LPG/NGL Pipeline and Facilities Map for the United States and Canada

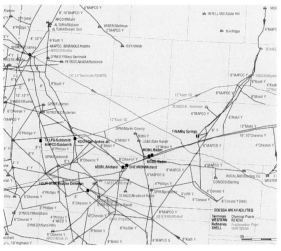

Modeling Land Use Change in Colorado Springs

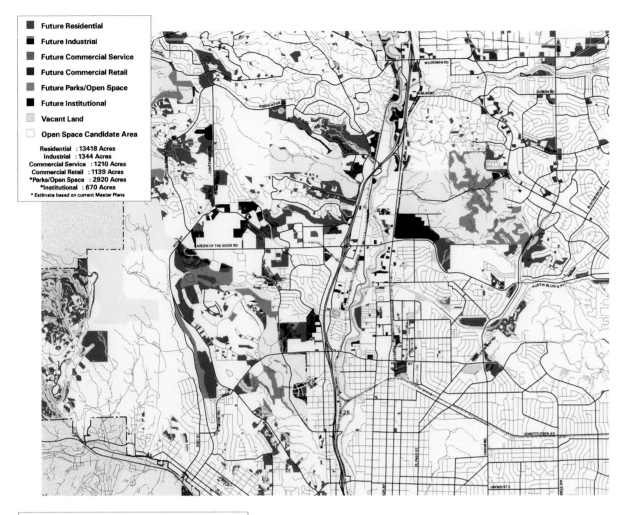

Legend:
- Future Residential
- Future Industrial
- Future Commercial Service
- Future Commercial Retail
- Future Parks/Open Space
- Future Institutional
- Vacant Land
- Open Space Candidate Area

Residential	: 13418 Acres
Industrial	: 1344 Acres
Commercial Service	: 1210 Acres
Commercial Retail	: 1139 Acres
*Parks/Open Space	: 2920 Acres
*Institutional	: 670 Acres

* Estimate based on current Master Plans

City of Colorado Springs
Colorado Springs, Colorado

By Phil Friesen and Ben Lydon

Contact
Phil Friesen
pfriesen@ci.colospgs.co.us

Software
ArcInfo 7.2.1 and ARC GRID
Hardware
Sun Ultra 1 workstation and
Windows NT workstation
Printer
HP DesignJet 755CM
Data Source(s)
City of Colorado Springs, Colorado
Springs Utilities, and El Paso County,
Colorado

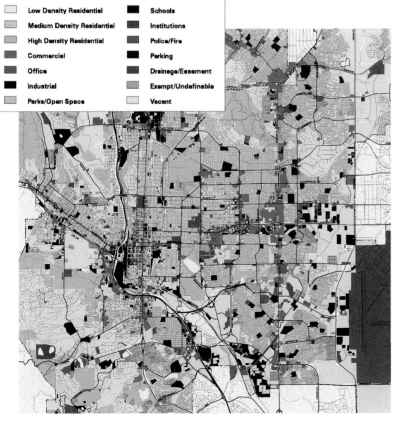

Legend:
- Low Density Residential
- Medium Density Residential
- High Density Residential
- Commercial
- Office
- Industrial
- Parks/Open Space
- Schools
- Institutions
- Police/Fire
- Parking
- Drainage/Easement
- Exempt/Undefinable
- Vacant

Faced with a rapidly expanding population and with 47 percent of its 186 square miles vacant, the city of Colorado Springs, Colorado, wanted to plan for its growth to the year 2020 with a strong focus on future land uses. The city's GIS enabled planners to model future land use patterns for the update of the city's comprehensive plan. ArcInfo and ARC GRID helped model four land use types—residential, retail, commercial services, and industrial.

A statistical technique, logistic regression, correlated existing land use with factors that have a significant influence on land use development. Land use could be predicted on the basis of known characteristics such as slope, transportation factors, or floodplains.

The resulting coefficients were applied to grid cells representing vacant land, yielding four probability surfaces. Cells were then allocated until specific targets, determined by market analysis, had been met. A base case scenario for the year 2020 and a set of alternative scenarios, reflecting potential changes in the development environment, were generated.

Regional Plan of Hame Region, Finland

Hameen Liitto, Regional Council of Hame
Hameenlinna, Hame, Finland

By Regional Council of Hame

Contact
Hannu Raittinen
hannu.raittinen@hameenliitto.fi

Software
ArcView GIS
Hardware
Sunlogix Pentium
Printer
Canon color laser copier 800 and
Color Pass 700-800
Data Source(s)
Regional plan data, basic map
(raster), lakes, and fields

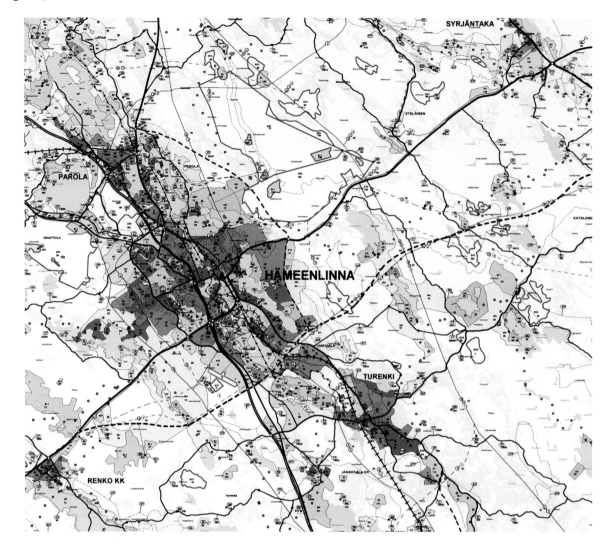

The Hame Regional Council is a joint municipal authority operating according to the principles of local self-governance with 16 municipalities. There are 20 regional councils in Finland. They operate as the authorities for regional development and as units for regional planning and look after regional interests. They articulate common regional needs and work to promote the material and cultural well being of their regions. Regional councils have statutory responsibility for regional development and planning and promote the interests of their regions through research, planning, and analyses. They are also responsible for implementing European Union regional development policy in their areas.

The basis for regional planning is the Building and Planning Act. The regional planning system in the Hame region encompasses the spatial plan, Hame 2020+, and the regional plan. The regional plan is the sanctioned plan that imposes certain legal obligations on the authorities and private landowners and ensures the long-range systematic development of land use. It guides local planning and other land use planning. The task of regional planning is to identify national land use objectives and trends within the area covered by the Regional Council of Hame.

The regional plan designates development, industry, recreation, conservation, traffic, technical services, and excavation areas. It incorporates the interests of both central and local government and harmonizes land use objectives with the aims of economic and cultural policy.

Issues are examined from the point of view of man, nature, and the built environment, with the principle of sustainable development playing a key role. The Regional Plan in the Hame region is based on four issues: (1) a convenient central location in southern Finland and in the Baltic Sea area; (2) versatile agricultural production and forestry and associated industries; (3) general and diverse industries; and (4) land use allocations for local and national training, tourism and recreation/conservation, and expansion and development of areas of outstanding natural beauty and historical value.

The Regional Plan of Hame was developed in five stages. The Ministry of the Environment ratified the latest stage in October 1998. The regional plan consists of the planning report and the map at 1:50,000 scale.

Analyzing Growth Plans

Cities in Central Puget Sound Region, 1998

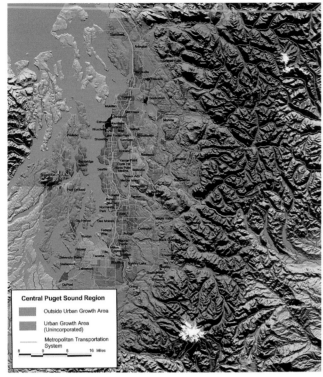

Central Puget Sound Region
- Outside Urban Growth Area
- Urban Growth Area (Unincorporated)
- Metropolitan Transportation System

Average Household Density, 1998

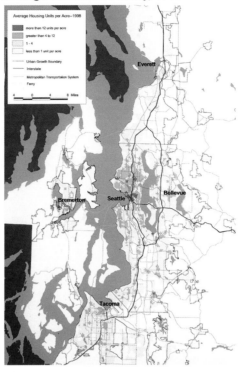

Average Housing Units per Acre—1998
- more than 12 units per acre
- greater than 4 to 12
- 1 - 4
- less than 1 unit per acre
- Urban Growth Boundary
- Interstate
- Metropolitan Transportation System
- Ferry

Puget Sound Regional Council
Seattle, Washington

By Eleanor Bell, Jay Clark, Carol Naito, Andrew Norton, and Nancy Tosta

Contact
Jay Clark
jclark@psrc.org

Software
ArcView GIS Spatial Analyst, ARC GRID, and ARC NETWORK™
Hardware
Sun Ultra 2 workstation
Printer
HP DesignJet 2500CP
Data Source(s)
U.S. Geological Survey digital elevation model, county and city land use plans, Washington Department of Ecology major water bodies, 1998 local building permits, and 1990 U.S. Census data

Washington's 1990 Growth Management Act (GMA) required that counties of more than 50,000 people and the cities within them develop comprehensive plans to manage growth. The plans address a variety of factors such as environmental protection, urban growth, commercial and industrial development, transportation, and housing.

The central Puget Sound region, with a population of 3.1 million people, has four counties—King, Kitsap, Pierce, and Snohomish—and 81 cities that have developed or, in the case of recent incorporation, are developing plans. A primary goal of all of these plans is to focus most growth in the region within the urban growth area (UGA).

The maps in this display depict various aspects of growth. City, county, and urban growth boundaries are displayed on the terrain of the region, and the 1998 density of housing is shown—the result of a 2.5-acre grid cell analysis. Currently, approximately 84 percent of the region's housing is inside the UGA. Buildout of comprehensive plans would yield roughly the same percentage of housing within the UGA. The difference between 1998 housing density and potential future density as a result of comprehensive planning is also depicted. By 2020 the region is expected to house approximately 4.3 million people.

The Regional Council used a variety of data sets and GIS tools to conduct this analysis. The 1990 U.S. census provided the baseline housing. Post-1990 housing estimates are derived from residential building and demolition permits.

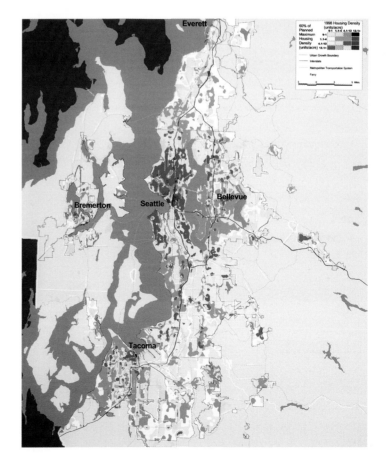

60% of Planned Maximum Housing Density (units/acre)

1998 Housing Density (units/acre)
- Urban Growth Boundary
- Interstate
- Metropolitan Transportation System
- Ferry

St. Johns County GIS Applications in Future Land Use and Open Space Preservation Analysis

St. Johns County
St. Augustine, Florida

*By Corey D. Bowens,
Mike J. Campbell, and
Tom A. Tibbitts*

Contact
Michael J. Campbell
gis@co.st-johns.fl.us

Software
ArcInfo 7.2.1, ArcPress, and ArcView GIS 3.1
Hardware
PC Network and Sun Enterprise 450 server
Printer
HP DesignJet 755CM
Data Source(s)
Federal Emergency Management Agency, St. Johns County GIS Program, Property Appraiser's Office, and Planning Department

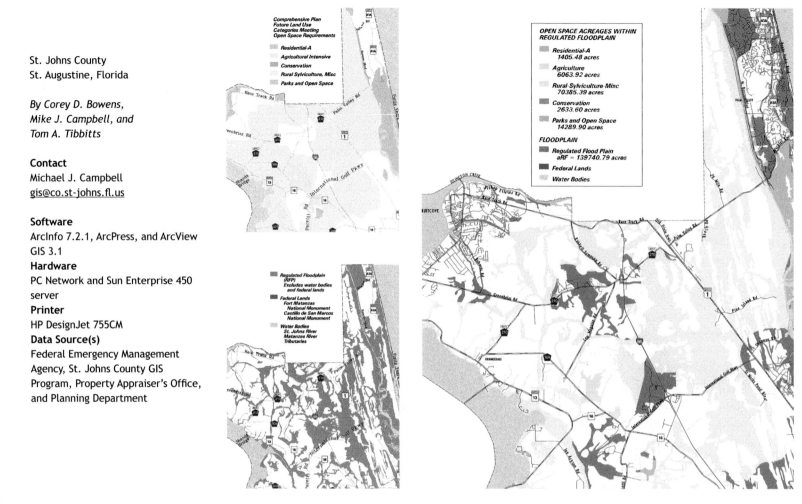

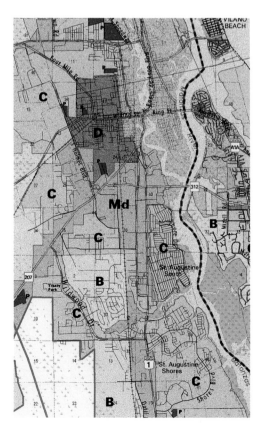

St. Johns County, Florida, is home to the city of St. Augustine, the oldest continuously occupied European settlement in the United States, as well as vast wetlands and forests, and it is a bedroom development to Jacksonville. The county needs accurate information to responsibly regulate growth.

The St. Johns County Future Land Use map is one in a series of maps used in the Comprehensive Plan Land Use Element, which is compiled and revised by the planning department. In an earlier form, the Future Land Use map showed general land use areas and was not site specific. This map has now taken a more detailed form, displaying future land use boundaries within the county's planning area by using natural, geographic, man-made, and parcel boundaries. This new map gives regional planners and other decision makers a spatial view of land use changes made in the county.

The Open Space Preservation Analysis map was created to calculate the total acreage of preserved open space within the regulatory floodplain of St. Johns County for the National Flood Insurance Program Community Rating System (CRS) application. Existing data layers of future land use, recreational sites, federal lands, and water bodies were intersected with the Q3 Flood Data from the Federal Emergency Management Agency to create a coverage of open space lands within the floodplain.

This effort dramatically increased the county's CRS points, which resulted in a lower classification number for the community. Residents of St. Johns County will benefit from having a lower classification number by qualifying for higher discounts on their insurance premiums.

Republic of Macedonia Spatial Plan, Land Use Projection

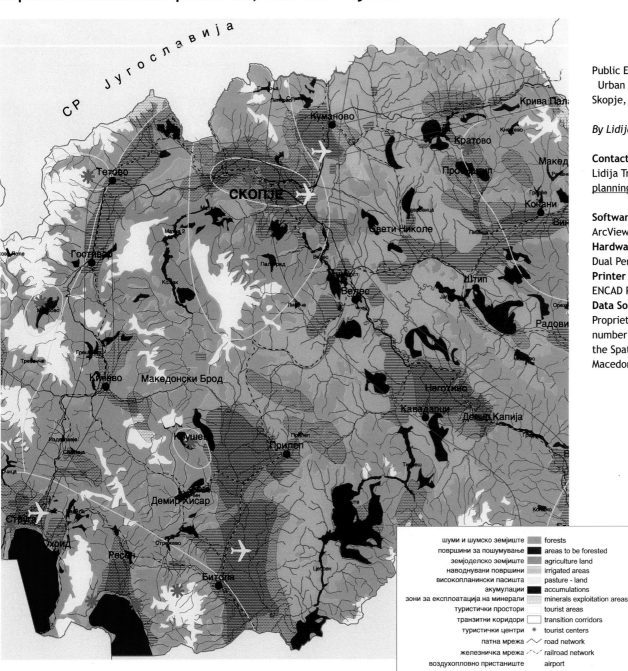

Public Enterprise for Spatial and
 Urban Planning
Skopje, Macedonia

By Lidija Trpenoska-Simonovik

Contact
Lidija Trpenoska-Simonovik
planning@unet.com.mk

Software
ArcView GIS 3.1
Hardware
Dual Pentium II workstation
Printer
ENCAD Pro50
Data Source(s)
Proprietary data derived from a
number of studies incorporated in
the Spatial Plan of Republic of
Macedonia

шуми и шумско земјиште	forests
површини за пошумување	areas to be forested
земјоделско земјиште	agriculture land
наводнувани површини	irrigated areas
високопланински пасишта	pasture - land
акумулации	accumulations
зони за експлоатација на минерали	minerals exploitation areas
туристички простори	tourist areas
транзитни коридори	transition corridors
туристички центри	tourist centers
патна мрежа	road network
железничка мрежа	railroad network
воздухопловно пристаниште	airport

The Spatial Plan of Macedonia is a government document that defines the spatial organization of the country. It is a complex development project articulating the goals and concepts for spatial development as well as the terms for their implementation. The project is long-term with anticipated completion in 2020.

The Land Use Projection map was produced by combining various themes and areas of interest. The concept was not only to present existing conditions, but also to make projections for 2020. Planners will be able to use this map as a tool for making planning decisions.

There are three land categories—forests, agricultural land, and pasture land. Existing areas for each category are shown as well as proposed areas to be forested or irrigated for agriculture. Areas with special environmental concerns, sections with mineral resources, tourist and recreation areas, and transportation corridors are also indicated.

Master Plan for the City of Guenzburg

Architektur + StadtPlanung
 Dr. Meister Ulm
Ulm, Germany

By Daniel Meister and Jorg Schaller

Contact
Daniel Meister
Fax: 49 731 619695

Software
ArcInfo, ArcView GIS, and ArcView
3D Analyst
Hardware
Pentium III 500
Printer
HP DesignJet 450C
Data Source(s)
Architektur + StadtPlanung
Dr. Meister Ulm planning data

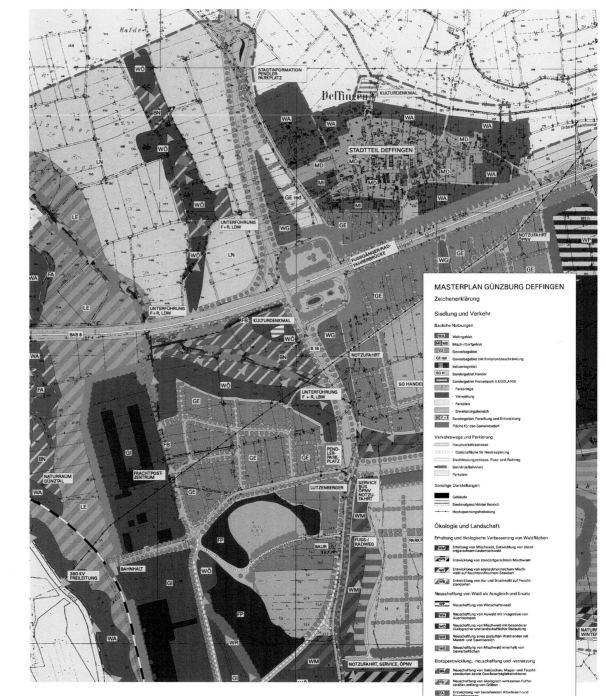

Guenzburg, a city in southern Germany, contracted with a planning team to develop a master plan for the implementation of the first German Legoland Park as well as an integrated concept for the urban development of the surrounding area. Problems such as traffic congestion, environmental constraints, and various requirements from the landowners, the public, and the administration were addressed in the master plan, which served as a discussion tool for all parties involved in the planning process.

An ArcInfo GIS map in 1:10,000 scale provided ease of data integration from different sources and data updating. Computer-aided design and GIS data from different engineering and government offices was georeferenced and overlaid with the planning data from the team. The results were delivered to the city in ArcView GIS shapefiles and attractive plot outputs. During the public hearings, three-dimensional maps were displayed for enhanced visualization.

Regional Planning in the Lausitz-Spreewald Region with ArcView GIS 3

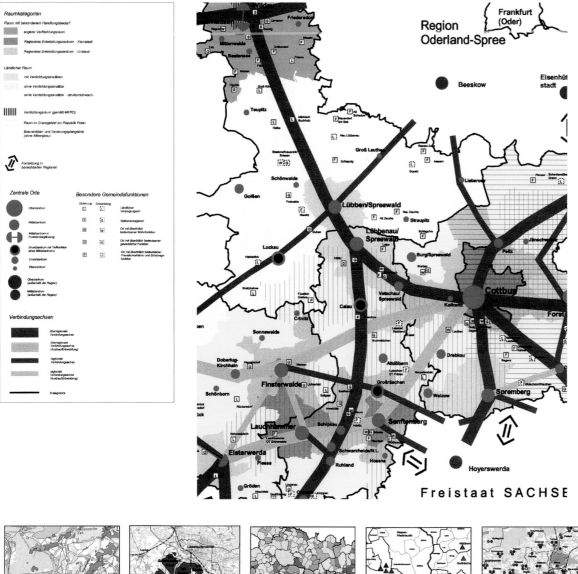

Regionale Planungsgemeinschaft
 Lausitz-Spreewald
Cottbus, Germany

By Jens Lochmann

Contact
Jens Lochmann
jenslochmann@aol.com

Software
ArcView GIS 3.0a, ArcPress for
ArcView 2, and ArcCAD® 11.4
Hardware
PC Pentium 200 and Windows NT 4.0
Printer
HP DesignJet 650C
Data Source(s)
LVermA Brandenburg and Regionale
Planungsgemeinschaft Lausitz-
Spreewald

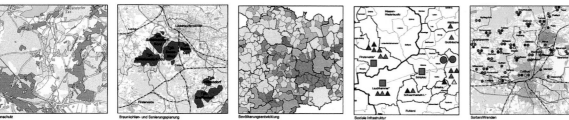

This poster shows the structure as well as the methods applied during the planning process for the development of the GIS-based regional plan of the Lausitz-Spreewald region, one of the five regions of the state of Brandenburg surrounding the city of Berlin.

The regional plan serves to protect and develop the ecological and economic resources of the region. It is an integrated multidisciplinary planning process, which balances the interests of the different communities.

The plan incorporates housing, infrastructure, open space development, and environmental protection. Specific topics, such as nature conservancy, water resources management, forestry, and agriculture, were surveyed, evaluated, and sorted according to spatial criteria of regional planning. The regional plan also indicates "areas of limited use" and "areas of priorities," which are defined by German planning law, and improved spatial planning, which is used to refine the statewide planning goals.

One map shows six steps involved in the method of visually overlaying the different categories of the regional plan. Graphics explain the different responsibilities with the regional planning agency, the planning process, and the legal basis for the plan.

The map of the final regional plan, an extensive written report with graphics, and additional thematic maps are also part of the regional plan.

Palo Verde 10-Mile Emergency Planning Area

Maricopa County Department of
Transportation
Phoenix, Arizona

By Karen Stewart

Contact
Karen Stewart
kstewart@mail.maricopa.gov

Software
ArcInfo 7.2.1
Hardware
Dell Precision Windows NT
workstation
Printer
HP DesignJet 650C and 750C+
Data Source(s)
Maricopa County Department of
Transportation coverages, aerial
photos, Maricopa County Flood
Control District, Arizona Public
Service

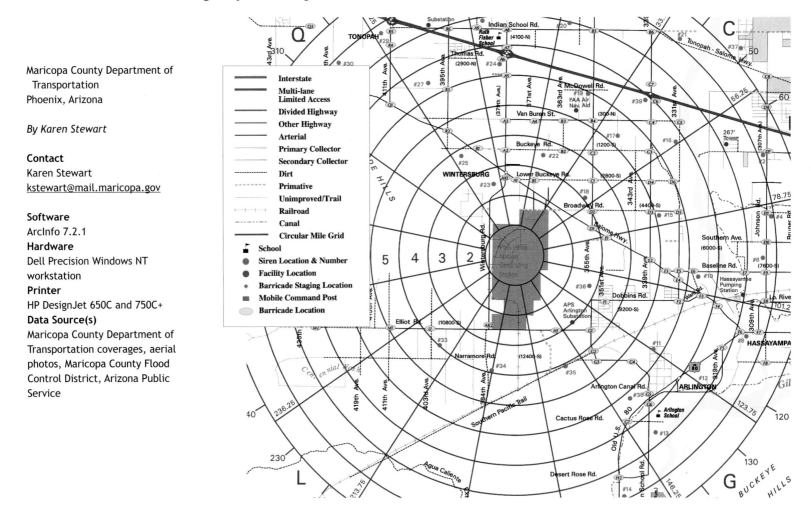

The Palo Verde 10-Mile Emergency Planning Area map is one of a series
that includes maps of various sizes covering an area ranging from a 10-mile
radius divided into sectors, out to a 50-mile radius. The Maricopa County
Department of Transportation GIS (MCDOT-GIS) created the maps for
the Maricopa County Department of Emergency Management for use
in the Emergency Operation Center (EOC) and by all assigned field
personnel.

The maps display all roads, emergency siren locations, schools, and
any possible locations for roadblocks. If an accident occurs, it is the
responsibility of the state of Arizona and Maricopa County to protect
the people living within the 10-mile radius of the Palo Verde Nuclear
Generating Station. Another responsibility of local government is to
ensure that no one within a 50-mile radius of the plant ingests
contaminated food or water. GIS helps to locate people, animals,
agriculture sites, and all the related data required to respond to an
emergency in the area.

These maps are provided to numerous agencies including federal, state,
county, and local governments; the Arizona Public Service; private
agencies; and numerous volunteers. By providing the same map and data
to all the agencies, decision making has become more effective.

Through the cooperative efforts of MCDOT-GIS, Maricopa County
Flood Control-GIS, and other county departments, the maps specifically
dealing with the nuclear plant evacuation plan have been developed into
an ArcView GIS application for Emergency Management to use in their
EOC. This application is very diversified and user-friendly. It uses all
the data in the static paper maps and provides the user with additional
spontaneous real-time data to heighten the level of analysis.

The application displays emergency response resources and easily organizes
events as they happen. It enables EOC personnel to interactively place or
remove barricades and deploy sheriff's deputies to locations. They also can
display contamination plumes, aerial photography, U.S. Geological Survey
maps, and population data; zoom to an address or geographic feature; get
current weather data; and send JPEG files via the Internet.

The application has a range of uses from managing floods and hazardous
spills to managing large events. Additionally, it continues to meet the
needs for evacuation planning by dealing not only with incidents specific
to the generating station, but also with any type of evacuation. This
application, and the cooperative effort by which it was created, is just one
more way GIS is improving government and providing a higher level of
customer service to the public.

Guide to Natural Hazards in the Portland, Oregon, Metropolitan Area

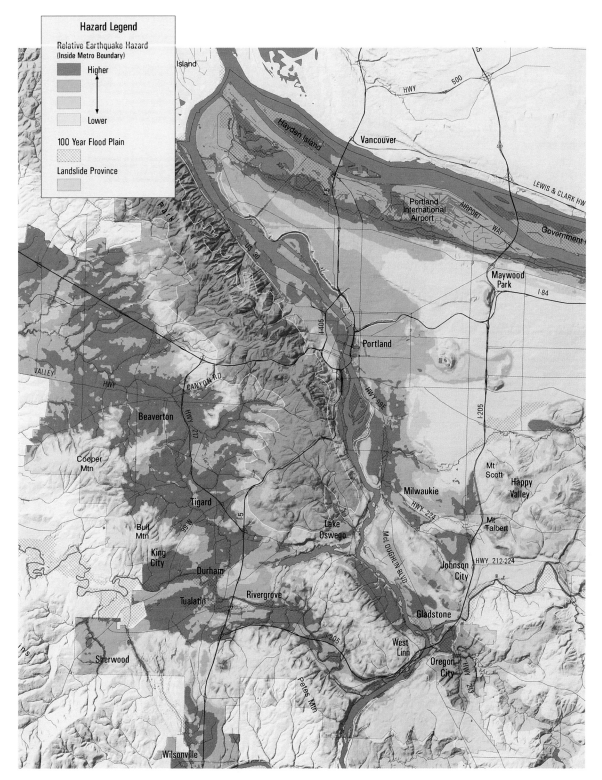

Hazard Legend

Relative Earthquake Hazard
(Inside Metro Boundary)

Higher

Lower

100 Year Flood Plain

Landslide Province

Metro
Portland, Oregon

*By Steven K. Erickson,
Susan Gemmell, and
Michael McGuire*

Contact
Steve Erickson
erickson@metro.dst.or.us

Software
ArcInfo 7.2 and
PageMaker Version 6.5
Hardware
PC and Macintosh G3
Printer
Offset
Data Source(s)
Metro

97 | public safety

This poster/guide provides an overview of natural hazards in the Portland metropolitan region as well as information on how to prepare for them. The hazards illustrated here include earthquakes, floods, and landslides. The front of the map shows the hazards and those areas most likely to experience damage from a particular hazard. The back provides information on how citizens can prepare to prevent or reduce the damage from such hazards.

Station Location/Resource Allocation Study: Road Travel Distances from Selected Montgomery County Fire Stations

Montgomery County
Rockville, Maryland

By Tim Taormino

Contact
Apollo Teng
apollo.teng@co.mo.md.us

Software
ArcInfo 7.1, ArcView GIS 3.1a, and
ArcView Network Analyst
Hardware
Sun UNIX workstation and
Windows NT workstation
Printer
HP DesignJet 3500CP
Data Source(s)
Montgomery County departments

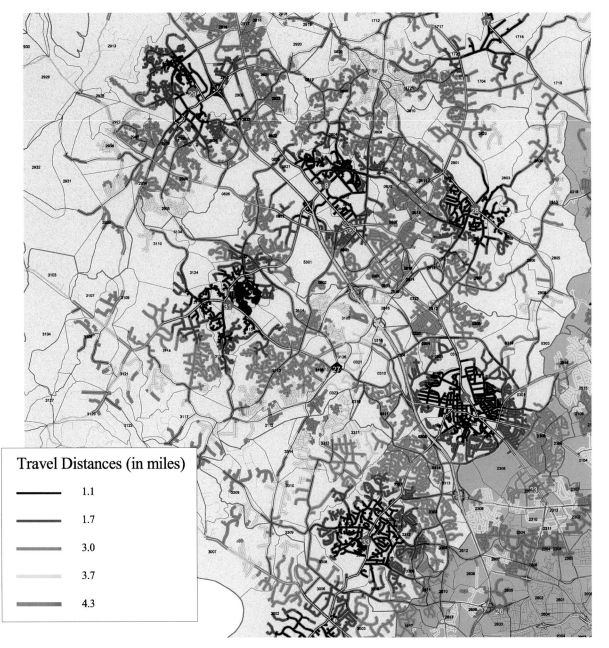

Travel Distances (in miles)

——	1.1
——	1.7
——	3.0
——	3.7
——	4.3

The Montgomery County Fire and Rescue Service (MCFRS) commissioned an in-depth study of the Clarksburg area in northwestern Montgomery County to examine population coverage within critical response times and to determine appropriate future station locations. This work was a follow-up to its preliminary study of rescue squad response coverage and in anticipation of the approval of a master plan covering more than 2,000 new residential land parcels in this area.

Previous station locations were based on historic precedent or the availability of public land, without regard to providing the best response to the most people.

The map shows the coverage of the current population from current station locations. The goal was to cover 90 percent of the population within five minutes' travel time (3.0 miles at 35 miles per hour) and 50 percent of the population within seven minutes' travel time (4.3 miles at 35 miles per hour). The ArcView Network Analyst extension was used to determine service areas at the designated distance thresholds.

MCFRS anticipated placing one new station in the Clarksburg area to meet the new demand and address existing service deficiencies. But, after studying this regional map and other maps of more specific areas, MCFRS is now considering adding more than one new station and relocating some of the existing stations to provide better overall service to the rapidly expanding Clarksburg–Germantown–Potomac region.

Market Planning for Telecommunications Services

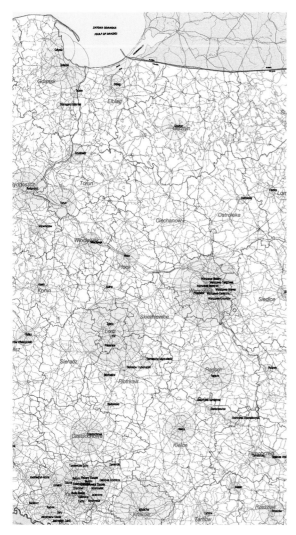

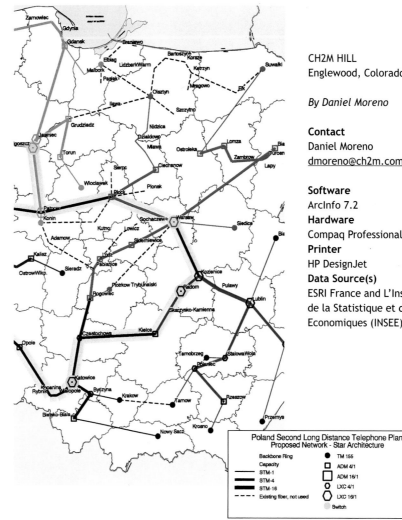

CH2M HILL
Englewood, Colorado

By Daniel Moreno

Contact
Daniel Moreno
dmoreno@ch2m.com

Software
ArcInfo 7.2
Hardware
Compaq Professional workstation
Printer
HP DesignJet
Data Source(s)
ESRI France and L'Institut National
de la Statistique et des Etudes
Economiques (INSEE)

Poland Second Long Distance Telephone Plan
Proposed Network - Star Architecture

Backbone Ring		TM 155
Capacity		ADM 4/1
STM-1		ADM 16/1
STM-4		LXC 4/1
STM-16		LXC 16/1
Existing fiber, not used		Switch

Telecommunications providers around the world are upgrading their systems to provide broadband cable services to residential and business customers. These services include traditional cable television, telephony, high-speed Internet data access, and two-way services such as video on demand.

In a fiercely competitive market, telecommunications providers must establish a business case for the huge investments required for plant upgrade and prioritize the rollout of upgrades by targeting the most profitable geographies.

In this study for a global telecommunications company, several demographic attributes were analyzed at the French commune level. This bivariate map depicts areas targeted for residential high-speed Internet access, namely those with high concentrations of upper-income families with children and heads of household in professional, executive employment.

Percent of Households in Commune

With Executive Head of Household

With Children

Automated Radio Design Support—Teligent RF Engineering

Teligent IT/Applications
Vienna, Virginia

By Jubal Harpster, Mike Ruth, and
Brian Sandrik

Contact
Mike Ruth
mike.ruth@teligent.com

Software
ArcView GIS, ArcView 3D Analyst, and
Avenue
Hardware
Dell Optiplex GX1P
Printer
Epson Stylus Color 1520
Data Source(s)
Aerial photogrammetry from 3D
Metric, ESRI ArcData, ESRI Streets-
USA, National Oceanic and
Atmospheric Administration, and U.S.
Geological Survey

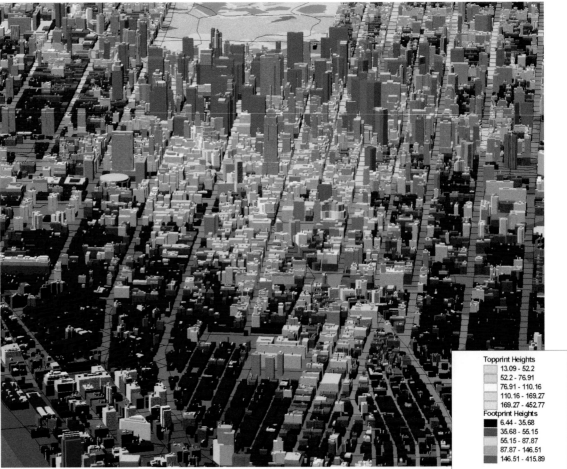

Topprint Heights
- 13.09 - 52.2
- 52.2 - 76.91
- 76.91 - 110.16
- 110.16 - 169.27
- 169.27 - 452.77

Footprint Heights
- 6.44 - 35.68
- 35.68 - 55.15
- 55.15 - 87.87
- 87.87 - 146.51
- 146.51 - 415.89

(Heights in meters from mean sea level)

Teligent, a global broadband communications company building fixed-wireless network links between commercial buildings, creates renderings like this simulated view of Manhattan to illustrate how three-dimensional geodata supports radio frequency (RF) engineering.

Teligent uses ArcView GIS software tools, including ArcView 3D Analyst and ArcView Spatial Analyst, to make the simulations, assign color and elevation values, and generate layouts from the three-dimensional geodata layers. Teligent also uses a customized set of software wizards—RF analysis tools written in Avenue™ language—to conduct RF engineering design and analysis on the three-dimensional geodata.

In many cities, fewer than 5 percent of commercial buildings are served by underground fiber optics. Fixed-wireless technology called local multipoint distribution systems (LMDSs) enables Teligent to supply big bandwidth to customers in buildings with or without fiber networks.

The network design requires an unobstructed line of sight between the transmitting and receiving antennae, which are mounted on commercial building rooftops. Modeling the many possible lines of sight and potential interference requires an accurate and detailed three-dimensional geodatabase. The modeled lines of sight support design and analysis of

the radio communications network on the computer desktop. The ability to model and simulate many RF propagation options in the computer reduces the cost of repeat rooftop visits for direct measurement and visualization of the network design in the field.

The Manhattan rendering uses color assignment and elevation extrusion to produce this realistic simulation of the skyline of Manhattan. The buildings are represented in the database by planar polygons with elevation attribution. Footprints and top-prints are symbolized in colors representing elevation. In this image, the polygons are also extruded by ArcView 3D Analyst.

With this mapping technology, Teligent can design the radio network, evaluate alternative configurations efficiently, and model the radio propagation characteristics analytically. The three-dimensional geodatabase also supports Federal Communication Commission application filings and network implementation. As Teligent's corporate integration activities continue to develop, the three-dimensional geodata will facilitate marketing, sales, and leasing activities and contribute to the entire life cycle of business, from initial market planning through design, implementation, and network operations.

New York CitiMap®

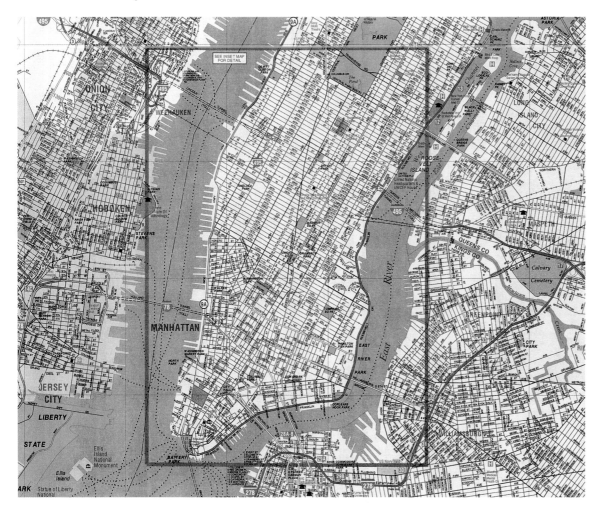

AAA
Heathrow, Florida

By AAA Map Production Team

Contact
Miguel Garriga
mgarriga@national.aaa.com

Software
ArcInfo 6.2.1, ArcStorm™, Maplex,
and Informix
Hardware
Sun Ultra Enterprise 4000 server
Printer
Lehigh Press
Data Source(s)
AAA, Navigation Technologies
Corporation, and Geographic Data
Technology, Inc.

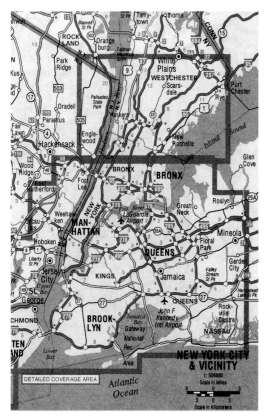

AAA has been producing maps since 1911. AAA GIS/Cartography incorporates traditional cartographic skills, automated production processes, and a multilayered GIS database to produce more than 44 million road maps annually for the association's 42 million members.

GIS cartographers extract geographic and road data from the AAA GIS database using tools developed in-house by AAA's GIS development and support team. All AAA GIS maps contain proprietary information supplied by the AAA Travel Information Department and are designed to meet members' touring needs.

The New York CitiMap is part of the AAA CitiMap® series. At ESRI's 1999 International User Conference, it received a first place award for best cartographic design in the Single Map Product category.

The map shows various views or insets of the New York City area. All views are drawn from features in a master database. Annotation was generated from Maplex, and all cartographic work was done with ArcInfo.

The AAA GIS database contains more than 50 gigabytes of data. The master GIS database supports cartography and other electronic applications. Data is updated only in the master database, and features are propagated to products through selection filters. The Maplex text placement is done within the standard map production work flow. On-the-fly projection (Mapprojection) is available for individual map views.

Bethlehem Tourist Map (Year 2000 Edition)

Good Shepherd Engineering and
Computing
Bethlehem, Palestine

*By Rana Halaby, Maher Owawi,
Khaled Ramadan, George Rock,
Amjad Yaghmour, Michael Younan,
Sana Younan, and Rima Zawahreh*

Contact
Michael Younan
gse@palnet.com

Software
ArcInfo 7.2.1, ArcView GIS 3.1,
ArcCAD, AutoCAD, and CADOverlay
Hardware
Windows NT workstations
Printer
HP DesignJet 3500CP
Data Source(s)
Orthophotography, photogrammetry,
cadastral data, municipal data, and
on-site data collection

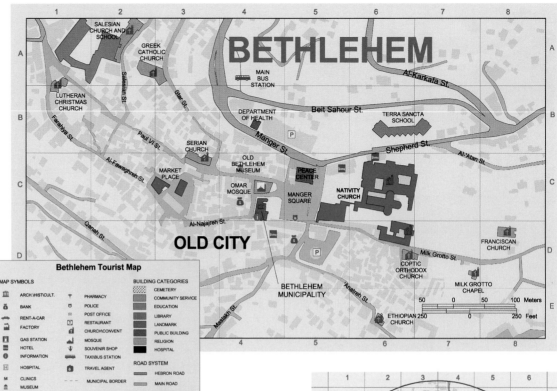

Good Shepherd Engineering and Computing (GSE) perceived a need for a tourist map of
Bethlehem as the year 2000 approached and the city prepared for an influx of visitors for New
Year's celebrations. The Bethlehem Tourist map is the first of its kind in Palestine produced by
GSE using GIS and computer-aided design for data entry, integration, analysis, and cartography.
An ArcView GIS EPS file was exported directly to press.

This two-sided map (scale 1:10,000) is a street map featuring the Bethlehem area that includes five
Palestinian municipalities—Bethlehem, Beit Sahour (Shepherds Field), Beit Jala, Al Khader, and Ad
Duha. It also includes Artas Village Council and three refugee camps. Detailed maps at 1:3,500
scale depict the Old City centers.

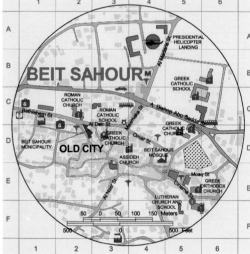

GSE designed this map to serve both tourists and local citizens. Tourists can navigate their way
to all attractions and services such as religious and historical sites, recommended tours and trails,
hotels, and restaurants—all clearly shown with symbols on the map. Local residents can use the
map to locate street details, government sites, and public buildings such as schools, banks, and
hospitals.

The other side of the map includes the Bethlehem Regional map, In the Footsteps of Jesus map,
a street and categories index, an intertown distances table, photographs of tourist attractions, and
a brief description of each site.

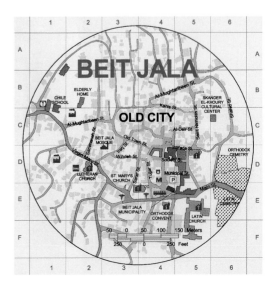

Minister Mitri Abu Aita, the Minister of Tourism and Antiquities of the Palestinian National
Authority, authenticated this map and wrote the introduction. The tourism ministry ordered
10,000 copies of the map and distributed them during Christmas and New Year's festivities in
Bethlehem. The Bethlehem Tourist map will be updated and published annually for worldwide
distribution.

Bike There!

Metro
Portland, Oregon

*By Bill Barber, Skye Brigner,
Sue Gemmell, Dave Petterson, and
J.O. Price*

Contact
Dave Petterson
petterd@metro.dst.or.us

Software
ArcInfo 7.2.1, Adobe Illustrator
Version 8, and PageMaker Version 6.5
Hardware
Pentium II Windows NT workstation
and Power Mac 7500
Printer
Offset printed
Data Source(s)
Metro

This map indicates bike route suitability classifications for the Portland metropolitan area. It was created to support transportation alternatives in the region and to promote continued improvement of Portland's outstanding network of bike routes and trails.

Whether you are going to work or school, to shop, or to visit friends, riding your bicycle is a great transportation choice for getting around. Providing an interconnected regional network of safe and convenient bikeways is one way Metro plans to meet the region's goal of increased transportation options.

Bike There! provides an up-to-date "snapshot" of bike land and multiuse paths in the metro region. In addition, the map rates selected thoroughfares where bicyclists share the road with motorists.

Streets shown on this map were chosen in consultation with cities, counties, citizen corps, recreation and commuter bicyclists, and transportation planners throughout the region.

In many cases, neighborhood streets with low traffic volumes and speed offer alternatives to some of the busier streets shown on the map. The shared roadway color-coded ratings can assist cyclists in finding streets to fit their individual needs.

The map is printed on waterproof paper and sold for six dollars. It features a section on safety tips and bike laws as well as areas where bicycles are prohibited and a list of references (e.g., recreational touring agencies, whom to call to report undesirable road conditions).

Metro serves 1.3 million people who live in Clackamas, Multnomah, and Washington counties and the 24 cities in the Portland metropolitan area. Metro provides transportation and land use planning services and oversees regional garbage disposal and recycling and waste reduction programs. Metro also manages regional parks and green spaces throughout the area.

Challenge and Response

Data Deja View
Northfield, Vermont

By Jim Mossman

Contact
Jim Mossman
jmossman@wcvt.com

Software
ArcInfo 8, ArcInfo 7.2.1, ArcPress,
ArcView GIS 3.1.1, ArcView Spatial
Analyst 1.1, ArcView 3D Analyst 1,
Lotus 1-2-3, and Visual FoxPro
Version 5.0
Hardware
Dell 410 and Windows NT workstation
Printer
Epson Stylus Color 3000
Data Source(s)
Vermont Agency of Transportation,
Vermont Center for Geographic
Information, and Data Deja View

In 1999, GIS was relatively new to the Vermont Agency of Transportation (VTrans) and was not fully utilized. The challenge was to produce maps that would be used as briefing tools by the top level of the organization and convince VTrans executives that GIS was worthy of their support.

The maps shown here were from a series produced for a legislative truck travel study committee. These maps offered new geographic perspectives on what had suddenly become a high visibility issue. Alternate route profiles and detailed views of involved villages rounded out the exhibits.

Other challenge maps were used to brief Governor Howard Dean on construction projects, compare roadway sufficiency changes, and show district facilities. The challenge was successfully met, as the maps garnered significant positive feedback resulting in VTrans acquiring the plotters and software necessary to produce these kinds of maps.

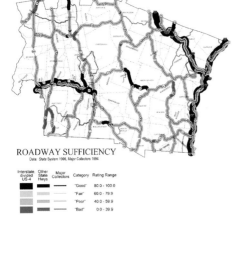

ROADWAY SUFFICIENCY
Data: State System 1996, Major Collectors 1994

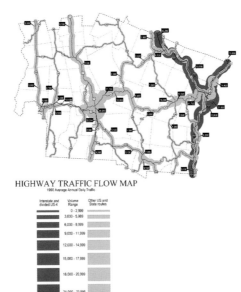

HIGHWAY TRAFFIC FLOW MAP
1996 Average Annual Daily Traffic

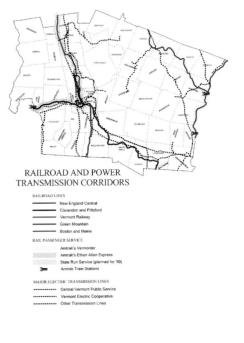

**RAILROAD AND POWER
TRANSMISSION CORRIDORS**

Status of Construction Work Program Projects—FY 1999 to FY 2005

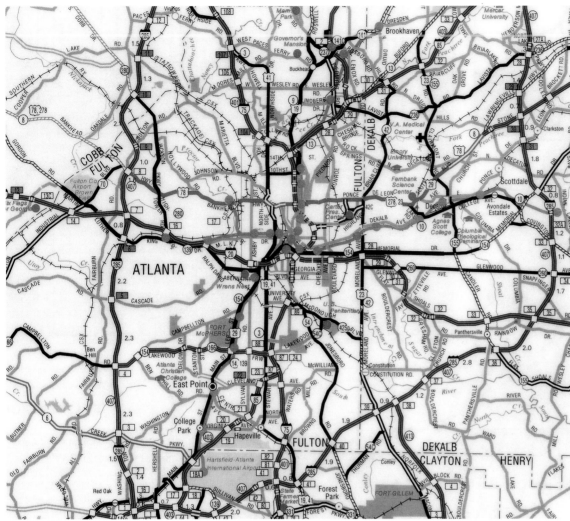

Georgia Department of Transportation
 Office of Information Services
 GIS Application Development Unit
Chamblee, Georgia

*By Steve Clement, Lynn Gulick,
Kum-Hoe Hwang, Tracey Leet,
Jack Martin, Todd Payne, and
Chunguang Zhang*

Contact
Chunguang Zhang
chunguang.zhang@dot.state.ga.us

Software
ArcInfo 7.1.2
Hardware
Digital 5000 AU
Printer
HP DesignJet 3500CP
Data Source(s)
Georgia Department of
Transportation (DOT) GIS Library
1:33,060-scale roads, routes, and
boundary coverages; annotation from
Georgia State Transportation Map;
Georgia DOT road characteristics
database; and linear events built
from dynamic-segment linkage to
Construction Work Program projects
database

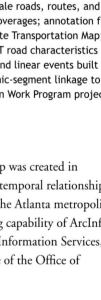

The Construction Work Program Projects map was created in
response to a need to visualize the spatial and temporal relationships
of the transportation improvement efforts in the Atlanta metropolitan
area. It was produced using the event mapping capability of ArcInfo,
the route system maintained by the Office of Information Services,
and the Construction Work Program database of the Office of
Programming.

The map shows the regional distribution of the transportation
improvement projects and helps with planning improved access to
transportation facilities in this rapidly growing urban center of the
southeastern United States.

Península de Cantera Project

Puerto Rico Department of
  Transportation and Public Works
Highway and Transportation Authority
San Juan, Puerto Rico

By Sonia Colon and Elba Torres

Contact
Sonia Colon
scolon@act.dtop.gov.pr

Software
ArcInfo 7.0.2, ArcView GIS 3.1, and
ArcView Image Analysis
Hardware
Pentium II Windows NT workstation
Printer
HP DesignJet 650C
Data Source(s)
Updated aerial photography

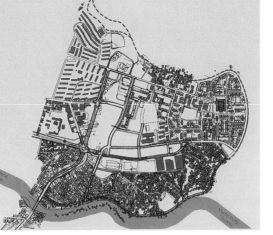

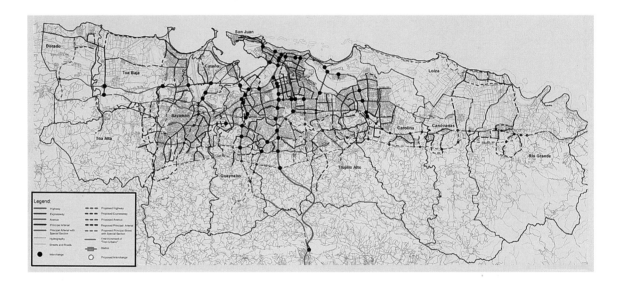

According to 1990 U.S. Census data, the Península
de Cantera is a low-income sector located in the
center of San Juan, the capital of Puerto Rico. This
project reflects the revised and adopted changes to the
San Juan Metropolitan Area Plan at the Península de
Cantera sector that were approved by the Planning
Board and the Highway and Transportation Authority.
These maps propose changes to the highway plan as
well as to the existing land use and zoning districts.

Five ArcView Views of Vermont in DataRoad

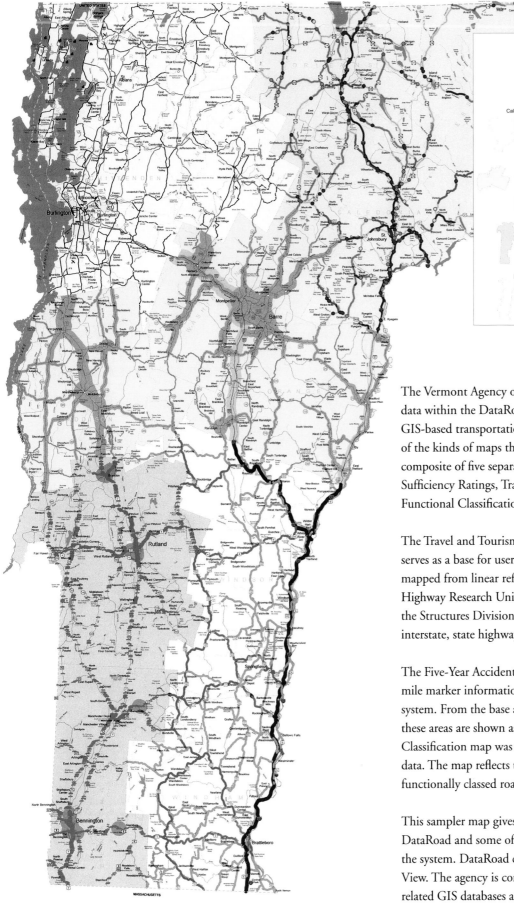

Travel and Tourism Map
Chittenden, Franklin, Grand Isle
and Lamoille Counties

A portion of Vermont's
Official State Map is
emulated in ArcView
using custom markers
for route signs and
other graphics.

Sufficiency Ratings
Caledonia, Essex and Orleans Counties

Highways	Bridges	
— Good	•	80.1 - 100.0
— Fair	•	60.1 - 80.0
— Poor	•	40.1 - 60.0
— Bad	•	0.0 - 40.0

Traffic Flow Map (AADTs)
Addison, Orange and Washington
Counties

A much simplified ArcView emulation of the existing
ArcInfo Traffic Flow Map application.

Interstate Highways
Other State System Roads

5-Year Accident History
Bennington and Rutland Counties

▴ Accident Site
● High Accident Location

Functional Classification Map
Windsor and Windham Counties

Interstate Highways
Freeways/Expressways
Principal Arterials
Minor Arterials
Collectors

Vermont Agency of Transportation
Montpelier, Vermont

By Johnathan Croft and Jim Mossman

Contact
Johnathan Croft
johnathan.croft@state.vt.us

Software
ArcView GIS 3.1
Hardware
Compaq Deskpro EN Pentium II
Printer
HP DesignJet 750C+
Data Source(s)
Vermont Agency of Transportation
and the Vermont Center for
Geographic Information

The Vermont Agency of Transportation (VAOT) developed this map to reflect data within the DataRoad system, which uses ArcView GIS as a tool to access GIS-based transportation data within the VAOT. It is intended as a sampler of the kinds of maps that can be produced at the agency. This map features a composite of five separate and unique maps including the Travel and Tourism, Sufficiency Ratings, Traffic Flow (AADTs), Five-Year Accident History, and Functional Classification maps.

The Travel and Tourism map emulates the official Vermont state map and serves as a base for users in cartographic production. Sufficiency ratings, mapped from linear referenced highway sufficiency data from the VAOT's Highway Research Unit, contrast with the bridge sufficiencies maintained by the Structures Division. The Traffic Flow map shows average daily traffic for interstate, state highways, and major collector roads.

The Five-Year Accident History map shows all the accidents with route and mile marker information for the last five years, mapped along the state highway system. From the base accident data, high accident locations are calculated, and these areas are shown as a background to the accident points. The Functional Classification map was derived from a code within the detailed road centerline data. The map reflects the current classification for the five highest categories of functionally classed roads within Windsor and Windham counties.

This sampler map gives an overview of some of the data available within DataRoad and some of the possible cartographic opportunities that exist within the system. DataRoad contains custom symbology developed by Data Deja View. The agency is continuing to build, maintain, and showcase transportation-related GIS databases and applications for its use and use by the GIS community.

Coastal Trunk Highway, Norway

Vestnorsk Plangruppe/Norconsult
Produced on behalf of Statens
  vegvesen Hordaland and Hordaland
  fylkeskommune
Bergen, Norway

By Trond Hollekim and Tom Potter

Contact
Thomas J. Potter
tom@vp.no

Software
ArcInfo, ArcView GIS, and EPPL7
Hardware
Dell Pentium III
Printer
HP DesignJet 750C+
Data Source(s)
Norwegian Map Administration and
Vestnorsk Plangruppe

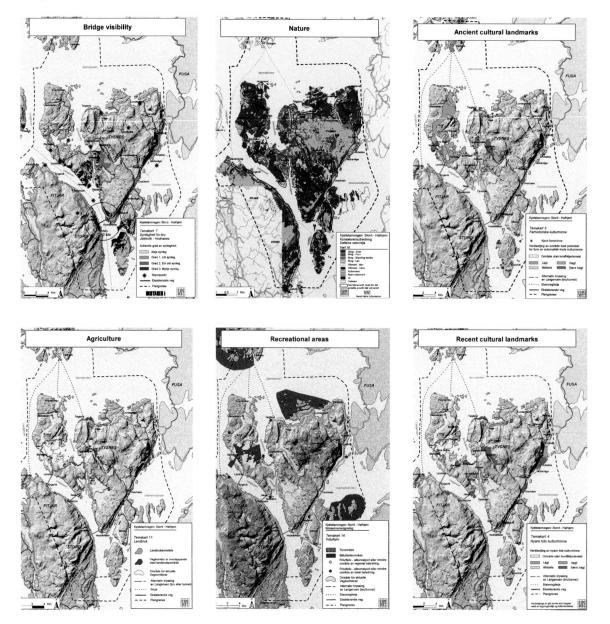

In 1995, the Norwegian Parliament considered the realignment of the Coastal Trunk Highway along the western coast of Norway. The thoroughfare could either continue with a long ferry connection between Stord and Os as it does today or a new connection to the island of Tysnes could be built, either as a bridge or tunnel, and connect to a shorter ferry to Os. In either case, the roads involved would be improved to a much higher standard than currently exists.

The County Highway Authority (Statens vegvesen Hordaland) and the County of Hordaland (Hordaland fylkeskommune) engaged the Vestnorsk Plangruppe (Bergen Norway) to analyze more than 2,000 possible alternative alignments over the island of Tysnes. These alternatives were then ranked according to the level of conflict with various area characteristics or environmental impacts such as agricultural, recreational, and cultural landmarks. ArcView GIS was used for all registration, analysis, and map production.

These maps highlight some of the various categories of environmental impact. Another larger map was produced to show all of the possible road alignments over Tysnes.

GIS Applications and Cartographic Products in Kentucky

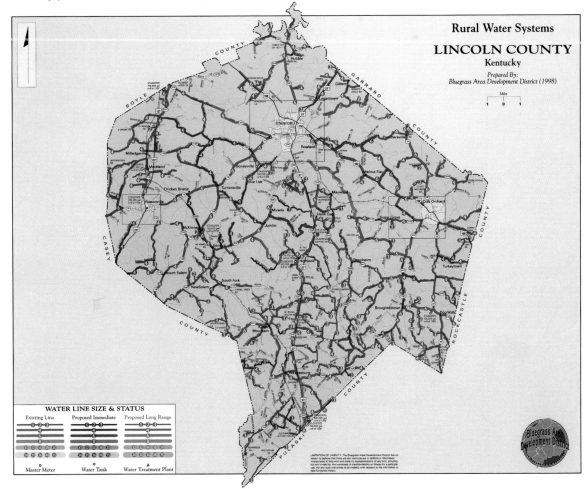

Bluegrass Area Development District
Lexington, Kentucky

By Rusty Anderson, Kent Anness, Brandon Jett, Jeff Levy, and Shane New

Contact
Shane New
shanen@bgadd.org

Software
ArcInfo 7.2.1, ARCPLOT, ARC Macro Language (AML), ArcView GIS 3.1, Adobe Acrobat, and Macromedia Freehand
Hardware
DEC Alpha Server, Windows NT workstations, and Macintosh 9600
Printer
HP DesignJet
Data Source(s)
Kentucky Natural Resources and Environmental Protection Cabinet, Kentucky's area development districts, Water Resource Development Commission, and Kentucky Transportation Cabinet

According to 1990 U.S. census figures, more than 275,000 Kentucky households were not on a public water system. This thematic map shows existing and proposed water service in Lincoln County, Kentucky. The data associated with this map was compiled for each county in the commonwealth as part of the governor's effort to get water service to all Kentuckians by the year 2020.

Data gathering and GIS work commenced on this project in early 1998, and Kentucky's area development districts (ADDs) completed GIS input of proposed water service information in June 1999.

The Bluegrass Area Development District GIS staff used ARCPLOT and ARC Macro Language (AML) to generate Adobe PostScript ASCII text files, which they used in Macromedia Freehand 8.0.1 to create the final cartographic product.

GIS and Water Resources Master Planning

Irvine Ranch Water District
Irvine, California

By Michael Hoolihan, John Robillard,
Felicia Sirchia, and Scott Williams

Contact
Michael Hoolihan
hoolihan@irwd.com

Software
ArcInfo 7.2.1, ARCPLOT, ArcView
GIS 3.1, and Haestad WaterCad
Hardware
IBM RS6000 and PC workstations
Printer
HP DesignJet 755CM
Data Source(s)
Irvine Ranch Water District and
Thomas Bros. maps

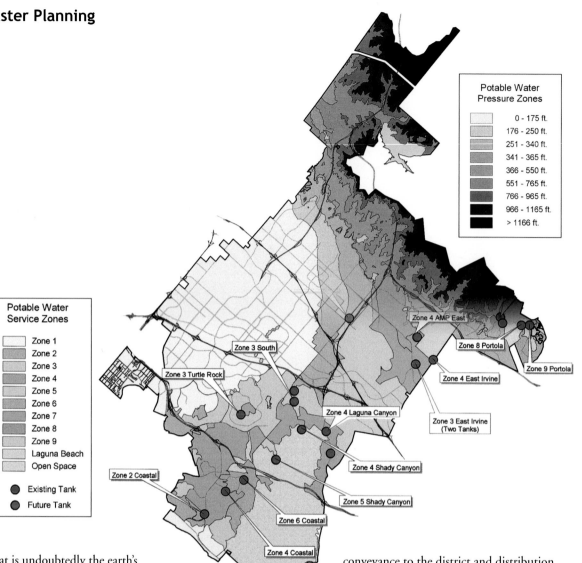

Potable Water
Pressure Zones

| 0 - 175 ft. |
| 176 - 250 ft. |
| 251 - 340 ft. |
| 341 - 365 ft. |
| 366 - 550 ft. |
| 551 - 765 ft. |
| 766 - 965 ft. |
| 966 - 1165 ft. |
| > 1166 ft. |

Potable Water
Service Zones

| Zone 1 |
| Zone 2 |
| Zone 3 |
| Zone 4 |
| Zone 5 |
| Zone 6 |
| Zone 7 |
| Zone 8 |
| Zone 9 |
| Laguna Beach |
| Open Space |

● Existing Tank
● Future Tank

Water is a simple word that describes what is undoubtedly the earth's most precious resource. But, the simple act of turning on a tap involves a complex system that starts with a comprehensive planning process in order to route it to users and ensure quality. At the Irvine Ranch Water District (IRWD) that planning process begins with the Water Resources Master Plan.

This series of maps gives examples of how GIS has enabled IRWD water resource planners to accurately estimate future water needs, plan for its conveyance to the district and distribution to end users, and precisely phase and budget capital projects.

GIS was used in every phase of the Water Resources Master Plan, and the final document contains more than 30 maps created directly from GIS and a number of other figures generated using GIS analysis. The entire Water Resources Master Plan is available for viewing at www.irwd.com.

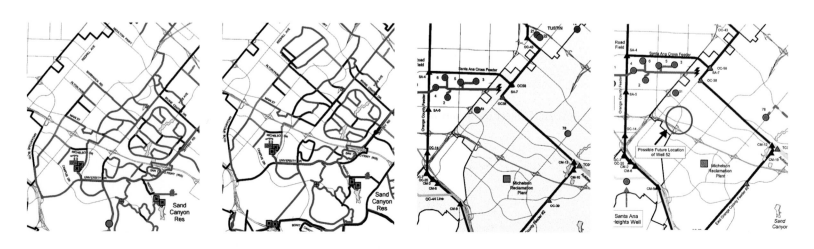

Environmentally Sensitive Features in Karst Geology Area

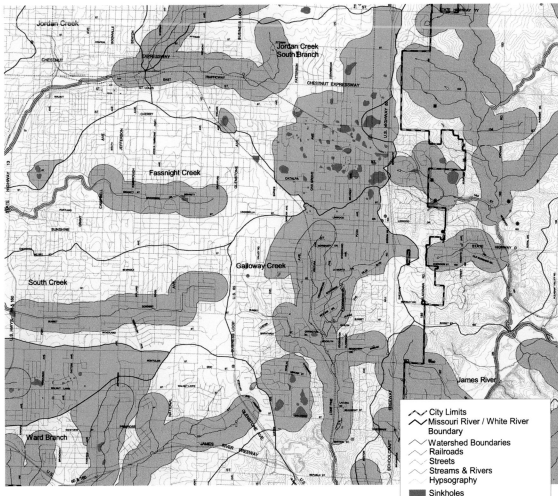

City of Springfield
Springfield, Missouri

By Kim Brown, Ken Carr,
Brad Chandler, Wendell Farrand,
Mike Fonner, Mike Giles,
John Hardin, Rich Miller, Mike Rude,
Gary Smith, Doug Thomas, and
Steve Tomlins

Contact
Mike Fonner
mike_fonner@ci.springfield.mo.us

Software
ArcInfo 7.2.1 and ArcView GIS 3.1
Hardware
HP UNIX
Printer
HP DesignJet 755C+

Legend:
- City Limits
- Missouri River / White River Boundary
- Watershed Boundaries
- Railroads
- Streets
- Streams & Rivers
- Hypsography
- Sinkholes
- ✳ Springs
- Missouri Dept of Natural Resources Special Erosion Control Required
- Losing Stream 1000' Buffer Permanent Stream 100' Buffer
- Sinkhole Interior Drainage Area

A karst terrain is one in which the dissolution of limestone, dolomite, gypsum, or marble by water plays a major role in land erosion. Karst areas are often characterized by a large numbers of caves, springs, sinkholes, and surface streams that disappear underground. All of these different types of water conduits contribute to the subsurface recharge and allow pollutants to enter the groundwater system.

Nine maps were created for this project to show the steps taken to identify a problem, analyze the problem, and come up with a solution. The problem was to establish a 10-year construction schedule to remove existing on-site wastewater treatment systems by providing sanitary sewers to the entire city of Springfield. An analysis determined those regions of the city where on-site wastewater discharge has the highest probability of adversely affecting the groundwater supply. The solution was to create an environmental impact index to establish a prioritized 10-year schedule for $20,000,000 of sewer construction to improve and protect the quality of surface and groundwater.

Cleveland Easterly Combined Sewer Outfall Project

Northeast Ohio Regional Sewer
 District
Cleveland, Ohio

*By Jeff Amero, Metcalf & Eddy, and
Joe Gautsch, CH2M HILL,*

Contact
Betsy Yingling
yinglingb@neorsd.org

Software
ArcInfo 7.1.1 and ArcInfo 7.2.1
Hardware
Digital UNIX Alpha and Sun Ultra 1
Printer
HP DesignJet 650C and 2500CP
Data Source(s)
Cuyahoga County Planning and
Northeast Ohio Regional Sewer
District

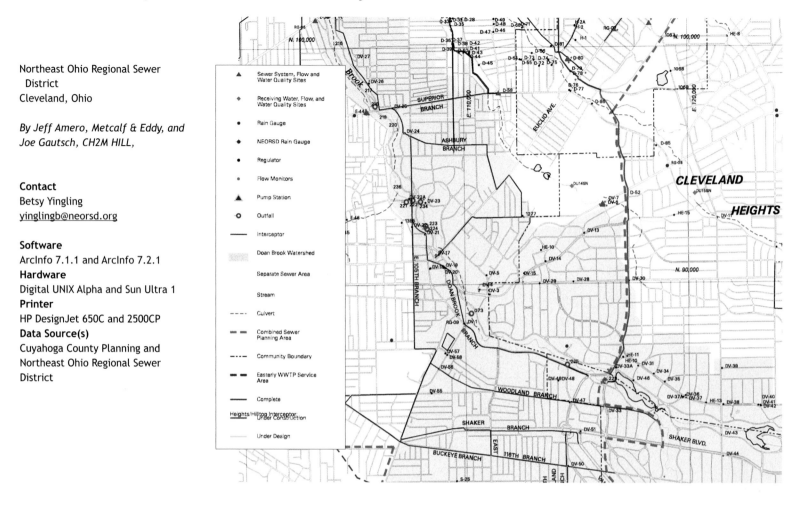

The Easterly Combined Sewer Overflow (CSO) Facilities Plan in
Cleveland uses GIS as the basis for data storage, retrieval, analysis, and
display of information about the collection and treatment of combined
sewer (wastewater and storm water) flows.

The information displayed on this map was compiled from Northeast
Ohio Regional Sewer District collection system maps and combined with
the various hard-copy and computer-aided design file maps from more
than 10 suburban municipalities. The effort standardized the data from
various sources into a detailed database design that included combined
sewer interceptors, separate sewers (sanitary and storm), and stream ways.
The project was both challenging and rewarding.

The plot displays the combined sewer, separate sewer, and tributary areas
with significant nodes selected and an upstream trace performed. Each line
is assigned to the node from which it was traced. The basins are derived
by performing Linegrid and Euclidean Growth commands on the traced
sewers and Eliminate on the resulting polygons.

The significant nodes are the treatment plants, pump stations, flow
slits, regulators, and monitoring stations. By combining population and
labor statistics and industrial flows for each basin, local area flows can

be estimated and calibrated. The local flows can then be accumulated
through the collection system. The model results are useful in assigning
maintenance priorities and assisting in the capital improvement planning
process.

The sewershed generation shown on the Easterly CSO and Doan Brook
Watershed Area map is part of a wastewater flow estimation model that
was developed for the city of Los Angeles in the early 1990s. It has since
been implemented in work for a number of other cities in the United
States.

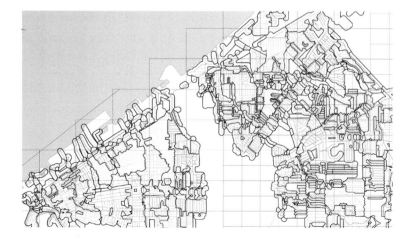

Sanitation Facilities Section Map

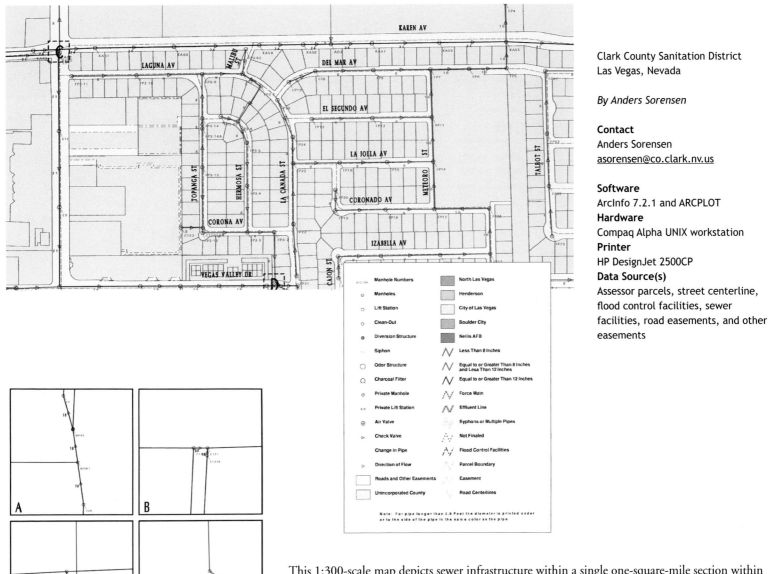

Clark County Sanitation District
Las Vegas, Nevada

By Anders Sorensen

Contact
Anders Sorensen
asorensen@co.clark.nv.us

Software
ArcInfo 7.2.1 and ARCPLOT
Hardware
Compaq Alpha UNIX workstation
Printer
HP DesignJet 2500CP
Data Source(s)
Assessor parcels, street centerline,
flood control facilities, sewer
facilities, road easements, and other
easements

This 1:300-scale map depicts sewer infrastructure within a single one-square-mile section within a township and range. A series of these maps covers the entire sewer system.

The map shows sewer pipes and manholes. Pipes are color-coded by pipe diameter with the diameters grouped into three categories. Different line symbols are used to display special types of pipes such as force mains and effluent lines. Point features are also displayed using different symbols to show their function—whether they are manholes or lift stations. Arrows are printed on pipes to show the direction of flow, and manhole numbers are printed by each manhole to identify it.

The sewer data is displayed against a background showing land parcels, flood channel easements, and street centerlines. Jurisdictional boundaries are also shown.

Sanitation district personnel use these maps in the field to locate lines and manholes, and in the office the maps support planning operations. Copies of the maps are also distributed to the public upon request.

Strategic Sampling Manholes in Jeffersontown, Kentucky

Louisville/Jefferson County
Information Consortium (LOJIC)
Louisville, Kentucky

By Joseph Wagner

Contact
Julia Muller
muller@msdlouky.org

Software
ArcView GIS 3.1
Hardware
Sun Ultra 1 workstations
Printer
HP DesignJet 750C+
Data Source(s)
LOJIC

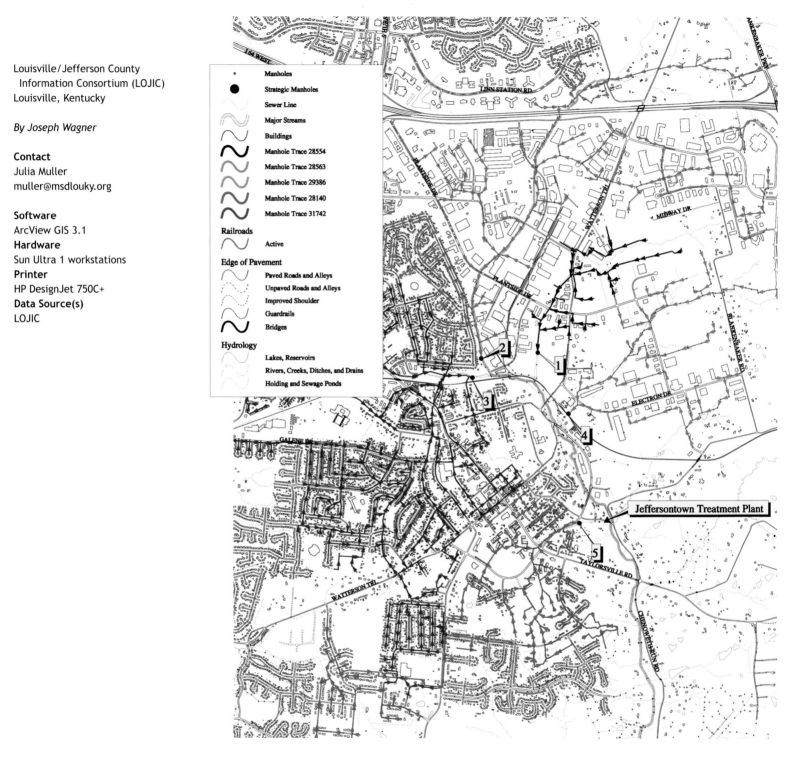

The Louisville and Jefferson County Metropolitan Sewer District (MSD) has been given a $250,000 grant from the U.S. Environmental Protection Agency for the development of performance measures for the pretreatment program. In order to determine the sources of pollutant loading coming to the wastewater treatment plant, monitoring points were located at the mouths of each of the five sewer subbasins of the sewer collection system. Wastewater samples and flow data are collected from each point.

The mass loading data collected from these points is an important tool for appropriately targeting resources to address the most significant sources of pollution in the collection system.

Municipal Utilities, City of Redlands

1998 Water Usage

Service Map

Legend (1998 Water Usage):
- 1 - 170*
- 170 - 262
- 263 - 391
- 392 - 1005
- Parcel Boundary
- Redlands City Boundary

City of Redlands
 Municipal Utilities Department
Redlands, California

By Dan Borell and Jeff Letterman

Contact
Dan Borell
dborell@eee.org

Software
ArcInfo 7.2, ARCPLOT, and ARC Macro Language (AML)
Hardware
DEC UNIX Alpha workstations
Printer
HP DesignJet 1050C
Data Source(s)
City of Redlands, Municipal Utilities Department and San Bernardino County

Service Map Legend:
- Valve
- Closed Valve
- 2.5" x 4" w/6" Riser - Std F.H.
- 2.5" w/4" Riser - Jones
- 2.5" w/6" Riser - Built-Up
- 2-2.5" x 4" w/6" Riser - Std. F.H. w/2-2.5"
- 2-2.5" w/6" Riser - Corey
- Fitting
- Blow-Off
- Air Vacuum
- Service Node
- Corporation Stop
- Disconnect
- Water Main
- Fire/Service Lateral
- Parcel Boundary
- Street Centerline
- Atlas Map Boundary

The map 1998 Water Usage shows total 1998 water consumption for each individual parcel served by the Redlands Municipal Utilities Department (MUD). The map was produced by linking geographic (parcel) with tabular (water consumption) data and classifying the information into four ranges using the quintile classification method.

The water meters are grouped into "routes," which are read on a bimonthly basis. When an entire route is read, the data is downloaded into the utility billing database where it, in turn, is downloaded into a GIS. ArcInfo cursors and ARC Macro Language (AML) are used to step through the meter read records and compute annual consumption. The resulting information is then related to the parcel map coverage using the parcel/meter address, and the map is produced.

This map helps department staff make sense of more than 18,000 meters and 100,000 yearly meter read database records. It is also a valuable tool for facilities planning.

The Redlands MUD maintains a domestic water supply system consisting of 346.4 miles of pipeline, 6,033 valves, 2,465 fire hydrants, 18,765 meters, and 17 reservoirs. All of this information is stored in an ArcInfo map coverage and plotted on MUD service maps. Locations are positioned relative to parcel boundaries and street centerlines.

Each service map covers an area approximately 3,000 feet by 2,000 feet and represents a single map sheet from a service map index, which covers the entire city in addition to areas served outside the city limit. The service area comprises a total of 136 individual service maps.

Each of these 1:1,200-scale maps is fully annotated with county property assessor information, and features are labeled with attributes. These maps serve as critical reference information to city staff, the public, and contractors.

ESRI educational products cover topics related to geographic information science, GIS applications, and ESRI® technology. You can choose among instructor-led courses, Web-based courses, and self-study workbooks to find education solutions that fit your learning style and pocketbook. Visit www.esri.com/education for more information.

■ ESRI Special Editions

GIS for Everyone

Now everyone can create smart maps for school, work, home, or community action using a personal computer. Includes the ArcExplorer geographic data viewer and more than 500 megabytes of geographic data.
ISBN 1-879102-49-8

The ESRI Guide to GIS Analysis

From Andy Mitchell, the author of the best-selling GIS classic *Zeroing In: Geographic Information Systems at Work in the Community*, comes an important new book about how to do real analysis with a geographic information system. *The ESRI Guide to GIS Analysis, Volume 1: Geographic Patterns and Relationships* focuses on six of the most common geographic analysis tasks.
ISBN 1-879102-06-4

Modeling Our World

With this comprehensive guide and reference to GIS data modeling and to the new geodatabase model introduced with ArcInfo 8, you'll learn how to make the right decisions about modeling data, from database design and data capture to spatial analysis and visual presentation.
ISBN 1-879102-62-5

■ ESRI Software Workbooks

Understanding GIS: The ARC/INFO Method (UNIX/Windows NT version)

A hands-on introduction to geographic information system technology. Designed primarily for beginners, this classic text guides readers through a complete GIS project in 10 easy-to-follow lessons.
ISBN 1-879102-01-3

Understanding GIS: The ARC/INFO Method (PC version)
ISBN 1-879102-00-5

ARC Macro Language: Developing Menus and Macros with AML

ARC Macro Language (AML) software gives you the power to tailor Workstation ArcInfo software's geoprocessing operations to specific applications. This workbook teaches AML in the context of accomplishing practical Workstation ArcInfo tasks and presents both basic and advanced techniques.
ISBN 1-879102-18-8

Getting to Know ArcView GIS

A colorful, nontechnical introduction to GIS technology and ArcView GIS software, this workbook comes with a working ArcView GIS demonstration copy. Follow the book's scenario-based exercises or work through them using the CD and learn how to do your own ArcView GIS project.
ISBN 1-879102-46-3

Extending ArcView GIS

This sequel to the award-winning *Getting to Know ArcView GIS* is written for those who understand basic GIS concepts and are ready to extend the analytical power of the core ArcView GIS software. The book consists of short conceptual overviews followed by detailed exercises framed in the context of real problems.
ISBN 1-879102-05-6

ArcView GIS Means Business

Written for business professionals, this book is a behind-the-scenes look at how some of America's most successful companies have used desktop GIS technology. The book is loaded with full-color illustrations and comes with a trial copy of ArcView GIS software and a GIS tutorial.
ISBN 1-879102-51-X

Zeroing In: Geographic Information Systems at Work in the Community

In 12 "tales from the digital map age," this book shows how people use GIS in their daily jobs. An accessible and engaging introduction to GIS for anyone who deals with geographic information.
ISBN 1-879102-50-1

Serving Maps on the Internet

Take an insider's look at how today's forward-thinking organizations distribute map-based information via the Internet. Case studies cover a range of applications for ArcView Internet Map Server technology from ESRI. This book should interest anyone who wants to publish geospatial data on the World Wide Web.
ISBN 1-879102-52-8

Managing Natural Resources with GIS

Find out how GIS technology helps people design solutions to such pressing challenges as wildfires, urban blight, air and water degradation, species endangerment, disaster mitigation, coastline erosion, and public education. The experiences of public and private organizations provide real-world examples.
ISBN 1-879102-53-6

Enterprise GIS for Energy Companies

A volume of case studies showing how electric and gas utilities use geographic information systems to manage their facilities more cost effectively, find new market opportunities, and better serve their customers.
ISBN 1-879102-48-X

Transportation GIS

From monitoring rail systems and airplane noise levels, to making bus routes more efficient and improving roads, this book describes how geographic information system technology has emerged as the tool of choice for transportation planners.
ISBN 1-879102-47-1

GIS for Landscape Architects

From Karen Hanna, noted landscape architect and GIS pioneer, comes *GIS for Landscape Architects*. Through actual examples, you'll learn how landscape architects, land planners, and designers now rely on GIS to create visual frameworks within which spatial data and information are gathered, interpreted, manipulated, and shared.
ISBN 1-879102-64-1

GIS for Health Organizations

Health management is a rapidly developing field, where even slight shifts in policy affect the health care we receive. In this book, you'll see how physicians, public health officials, insurance providers, hospitals, epidemiologists, researchers, and HMO executives use GIS to focus resources to meet the needs of those in their care.
ISBN 1-879102-65-X

ESRI Press
380 New York Street
Redlands, California 92373-8100

ESRI Press publishes a growing number of GIS-related books. Ask for these books at your local bookstore or order by calling 1-800-447-9778. You can also shop online at www.esri.com/gisstore. Outside the United States, contact your local ESRI distributor.